WEEDS AND WORDS

WEEDS AND

The Etymology of the Scientific Names

Iowa State University Press / Ames

WORDS

of Weeds and Crops

ROBERT L. ZIMDAHL

Robert L. Zimdahl is professor, Department of Plant Pathology and Weed Science, Colorado State University, Fort Collins.

Composed and printed in the United States of America

First edition, 1989

Library of Congress Cataloging-in-Publication Data

Zimdahl, Robert L.
Weeds and words : the etymology of the scientific names of weeds and crops / Robert L. Zimdahl.—1st ed.
p. cm.
Bibliography: p.
Includes index.
ISBN 0-8138-0128-1
1. Weeds—Nomenclature. 2. Crops—Nomenclature. 3. Latin language, Medieval and modern—Technical Latin. 4. Latin language, Medieval and modern—Etymology. I. Title.
SB611.Z56 1989
632′.58′014—dc19 88-21803
CIP

CONTENTS

PREFACE

THIS book is about plants, that is, the names of plants, primarily weeds, that are familiar to people in many parts of the world. These plants, as with most plants, have at least two names: one or more common names and the scientific name.

I first became interested in the scientific names of plants when I encountered great resistance among students to learning what they regarded as useless, boring, and perhaps even nonsense words designed by professors to confuse them. The arguments against learning these words are manifold, as anyone who has tried to teach them knows. The first defense is that the names are difficult because they are in Latin, which after all is a dead language. Outside of the Roman Catholic Church, few speak it at all and knowing Latin certainly doesn't score many points with one's peers. Besides, the argument continues, common names are widely accepted and convey real meaning. One must grant that Latin is difficult, but difficulty should be dismissed as an objection not worthy of one engaged in higher education. Like most worthy goals, obtaining an education will not be achieved without some effort. To pursue the argument concerning Latin, it is admitted that Latin is dead, but therein lies one of its important advantages as a medium for naming things. A dead language does not evolve and assume new forms as daily usage modifies it and introduces variation. The rules are fixed, and while the language can be manipulated, it is not pliable as is a living language.

As opposed to common names, scientific names have a universal meaning. Those who know scientific names will be able to verify a plant's identity by reference to standard texts or will immediately know the plant being referred to. Thus, those who do not share the same native language can make use of an unchanging language to share information.

Although Latin is considered to be the language of scientific nomenclature, it is not exclusively so. Scientific names are derived from a vocabulary that is Latin in form and usually Latin or Greek in origin. Other peculiarities that make scientific nomenclature difficult are the frequent inclusion of personal names, latinized location names, and words based on other languages. There are rules for name creation, which have been accepted by taxonomists, that provide latitude for imagination and innovation but not license for their neglect.

It is the purpose of this book to help students and others overcome their difficulties with scientific names and to familiarize them with their origin. In many cases the etymology of the scientific name serves as an aid to identification and can make the whole process of learning more enjoyable.

While this book is about weeds, we must acknowledge that, for the most part, weeds are only important because they interfere with some human activity, particularly the most important human activity—producing food. Therefore, the etymology of some of the world's important food crops has been included.

Of the 1934 weed species compiled by the Weed Science Society of America (Composite list of weeds 1984), 228 weed species from 57 families are discussed. There is a sampling of weeds from the important weed genera, climatic regions, and crops of the world. Also discussed are 35 species of important food crops from 15 families. The purpose was not to be exhaustive but to be informative and a little provocative. The text for each species discussed includes the binomial scientific name with the authority, a pronunciation guide (with contrived phonetics; e.g., no long and short vowel symbols are used), the common English name, and the etymology of the binomial name with the language of origin when appropriate. Each etymology is followed by a narrative to help readers understand the relationship of the name to the plant. Where such a relationship does not exist, the origin of the name may be obscure. Both the weeds and the crops are also listed by plant family.

INTRODUCTION

WHEN I was young I wondered why my parents had named me Robert. As far as I knew there wasn't another Robert in the family, so it couldn't be considered a family name. My name didn't seem to be a popular one at the time either, for there really weren't many Roberts around. However, my best friend had almost the same name; he was called Robertson, which seemed a bit odd because he wasn't Robert's son at all. His father's name was Charles Bertram.

I learned that Webster's dictionary included a section giving the meanings of names. I didn't gain this information through any elegant detective work on my part. My parents had turned to the dictionary to find a name for my younger sister who was soon to arrive. I checked this source and found that Robert was of Germanic origin and meant "bright in fame." The Germanic origin was consistent with my father's lineage so their choice made sense to me and because school was a burden then, I seized on the bright part, assuming, I suppose, that the fame would take care of itself later on. The attribute of being bright because I had been given a name that carried the quality of brightness was really appealing, and I gained renewed faith in myself, in my academic ability, and in my parents' perceptive abilities.

It was much later, when the brightness hadn't shown with the expected luster, that I learned that names were chosen by whim, by sound, for family reasons, or by consulting the dictionary, not because their etymology (origin and meaning) matched an incipient personality. Names had little predictive value. It was a difficult lesson!

Upon reflection, I decided it was, after all, nice to have a name, even if it didn't hold great promise or have predictive potential. It was also very convenient for my friends, acquaintances, and enemies to refer to me by name. It was so much easier than having to describe me in some complex terminology that included my external morphology and perhaps even my internal epistemology. It would be difficult to greet or talk of other people if they didn't have names. It is also interesting to ponder how often our impression of someone we don't know is governed by our opinion of someone else we do know with the same name.

Plants also have names, which, when we know their origin, help us to identify them and to tell something about them. For example, Iris, the maidservant of Hera, the Queen of Olympus, descended to earth on her many-colored scarf, which, when arrayed across the sky, was as picturesque as a rainbow. We really don't know whether the flower was named for the goddess or the story grew from the flower—perhaps a little of each. The story of Narcissus is equally appealing. The name may have come from *narke,* the Greek word for stupor, and may be an allusion to the plant's narcotic power. Or the name could have come from the story of the young, handsome Narcissus who ignored the romantic overtures of the nymph Echo and displeased the goddess Venus. Thus, one day while he was wandering by a stream, Narcissus noticed his own handsome reflection in the water, and when he paused to gaze at it, Venus made him fall in love with it. He could not reach or touch his image, but lingered so long that he became the graceful flower we call Narcissus, forever leaning from the bank to view its watery image. The etymological derivation from *narke* may be the more accurate, but the mythological story is more appealing and more apt to be remembered.

The preceding examples show how complicated and interesting the etymology of all plant names may be. Although personal names for people and common names for plants may be given arbitrarily, the etymology of a scientific name will usually tell us a great deal about the plant and its characteristics. However, the etymology does not always make sense and may, in fact, be meaningless because the original name was wrong or the characteristics used to name the plant were not obvious or important. It is true that every conceivable characteristic, visible and hidden, has been used to name some plants, resulting in names that may now seem obscure because we do not see the plant in the same way as the one who named it. The author of the name may have used a particular attribute of size, shape, color, taste, location, or resemblance that is no longer obvious to us. In most cases, however, the etymology is a reliable guide to the plant and tells us a great deal about it.

Scientific names can be taken from almost any source and be composed in a very arbitrary manner. Without some knowledge of the plant, it is not possible to tell whether the etymology has real meaning or not. Even when a person has some specialized knowledge of plants, the names often can remain meaningless.

Lily and Narcissus are examples of what the plant taxonomist calls common names; that is, the name most people, including students, know. These names are easy to learn, but for many plants, especially weeds, they are not universally accepted. Many weeds have several common names that vary with region within one country and also between countries. In the United States we call one weed slender foxtail, but in the United Kingdom, agriculturalists, for whom this weed is a serious problem, call it blackgrass. The common, annual broadleaf species Kochia has at least six common names, including kochia, fireweed, and Mexican fireweed. Some people call *Amaranthus retroflexus* L. pigweed, while others call *Chenopodium album* L. pigweed. The first weed is indeed a pigweed and the accepted common name is redroot pigweed. The second is not a pigweed and is more commonly known as common lambsquarters. Redroot pigweed is a common name that is really meaningful; it has a red root and pigs do eat it. It is more difficult to make sense out of the common name common lambsquarters.

If we see "wort" in a common name, we can assume the plant, or a part of it, was once used for food or medicine because "wort" comes from the old English *wyrt* or *wurt,* meaning root or plant. Similarly, many common names have the word "bane" as part of the name. This comes from the old English *bana,* but the ultimate etymology is unknown. When used as part of a plant name, it means that the plant is capable of destroying life; thus, it was used especially for poisonous plants. Such uses and knowledge may have been lost or are no longer common and thus render the common name interesting but not helpful to understanding the plant's characteristics.

Although the perennial weed quackgrass is familiar to many

farmers and homeowners, the name really doesn't mean much when you think about it. It is a grass, but it doesn't quack. In Europe the common name is couch grass (pronounced cooch), which is close to but is not quack. Quack is likely an Americanization of the British couch. Couch is not related to the English couch (sofa) although it could be a reference to the dense growth of this perennial above and below ground. It was derived from the Anglo-Saxon *civice,* which means vivacious, surely a reference to the vitality and endurance of this aggressive weed.

The common name cattail makes immediate sense to the observer. Anyone who sees the cylindrical, brown stalk of female cattail flowers will not find it too difficult to see the resemblance to a cat's tail. The common name dandelion requires a little more imagination than most can muster. It is said that it comes from the French *dent de lion,* which was derived from the Latin *dens leonis,* because someone saw a resemblance between the serrated leaves or the ray flowers and a lion's teeth. Probably very few of us think of the dental apparatus of a lion when we see yellow dandelion flowers.

Both history and geography have contributed to naming jimsonweed. In 1676 Nathanial Bacon led a rebellion in Jamestown, Virginia, and British troops were deployed to quell the uprising known as Bacon's rebellion. The troops were forced to forage for food and ate some of the common weed that we now call jimsonweed; they suffered partial, but not permanent, paralysis and nausea. The common name is a corruption of Jamestown.

Common milkweed is easier to deal with than its forbidding scientific name *Asclepias speciosa* Torr. It has a milk-colored juice, which is more properly called a latex. In the past it was used in herbal medicine to enhance the passage of body fluids and to increase human milk flow. The problem with calling it milkweed is that there are many other plants that also have a milky latex and therefore the name is imprecise, albeit very descriptive.

Scouring rush is commonly called horsetail or Mormon tea. In its environment the summer form resembles a horse's tail and is reputed to have been used by the Mormons to brew a bitter tea during their trek westward across the United States. It also has a high silicon content, which gives it abrasive qualities; it can be used by the modern backpacker to scrub dirty pots.

Common names also can have origins that were once meaningful but whose meaning has been lost to the user. An example of a name that is of obscure origin to all but the plant historian or perhaps to a historian of religious practices is common St. johnswort. In northern temperate climates this plant blooms around the time of the liturgical feast of St. John the Baptist, June 24th. The mythology surrounding the plant says that if one collects it on that day it will give protection from evil spirits and lightning strikes. This seems an odd combination of abilities for one plant, but it makes a charming story, and it just may work. I actually collected it in Colorado on June 24th, and to the best of my knowledge I have not been plagued by an evil spirit and I know I have not been struck by lightning, yet.

There are many examples that illustrate how a common name does or does not describe a plant accurately. They are useful names that have a variety of origins and thus, meanings depending on where and how they originated. However, they do not serve as the lingua franca among plant scientists as does the scientific name.

The system of botanical nomenclature today is difficult but not as difficult as it once was. When people first started to describe and name things, they completely described them in Latin. These descriptions became the plant's name. It was a cumbersome system that became less and less workable as more and more was learned about the objects being named and the descriptions became longer; only a learned few could actually name a plant. Caspar Bauhin (1560–1624) devised a system whereby everything would be named in only two words. It remained for the Swedish botanist/naturalist Carl Linnaeus (1707–1778) to undertake the task of naming and classifying the whole living world. He refined Bauhin's system of using two words, and *Species Plantarum,* first published in 1753, is regarded as the beginning of the modern system of plant nomenclature. In the Linnaean system the first word of the two-part name is always the genus name and is always capitalized. The second word is the species epithet and is never capitalized (both names are often italicized). Readers interested in learning about these rules are referred to some of the references in this book, particularly those by Stearn (1973), Johnson (1971), and Jaeger (1959).

Jaeger (1959) lists ten subdivisions of plant names: (1) classical names and their compounds; (2) native names; (3) geographical names; (4) personal names; (5) names indicative of age; (6) names indicating size, form, color, or resemblance; (7) names indicating habit or habitat; (8) names based on special characteristics; (9) names of fanciful or poetic application; and (10) names founded on error. All these categories exist among the scientific names of the large group of plants we call weeds, and the reader will find several of each herein.

This elegant yet simple system that we use today provides a rich diversity of names, some of which are easy to pronounce, spell, and remember: *Avena fatua* L. (A′-v-na fat′-u-ah), wild oats. Others present difficulties in all three categories: *Dactyloctenium aegyptium* (L.) Willd. (Dak-tih-lock-ten′-e-um e-gyp′-t-um), or wild crowfootgrass; *Gutierrezia sarothrae* (Pursh) Britt. & Rusby. (Goo-t-air-e′-z-uh sah-ro′-thray), or broom snakeweed.

The L. after *Avena fatua* means that the plant was named by Linnaeus and has retained the name he gave it. In other cases Linnaeus gave the plant a name but the description or specific taxonomic relationships have been modified by subsequent scientists. For example, *Cynodon dactylon* (L.) Pers. was originally named by Linnaeus (denoted by the L.) and subsequently modified by the German botanist Christian Hendrik Persoon (1761–1836, denoted by the abbreviated name Pers.).

Although Linnaeus started the system of binomial nomenclature now used throughout the world, he could not have lived long enough to name all living things. Therefore, it is common to find plants that he did not name and where the names given by the original author have also been modified. *Astragalus bisulcatus* (Hook.) Gray, or two-grooved milkvetch, was originally named by William Jackson Hooker, a British botanist (1785–1865), and subsequently modified by the American botanist Asa Gray (1810–1888).

No scientific name is complete without inclusion of the author's name (authority), usually in abbreviated form, which allows one to trace names and plants through the botanical literature.

Weeds, whether known by their common or scientific name, are familiar to most people throughout the world, certainly to farmers. The lives of many of the world's farmers can be described as a constant battle against agricultural pests, with weeds often the most important among them. In many parts of the world the farmer is still a person with a hoe.

These ubiquitous, common, and bothersome plants have been described in terms of their habitat, their behavior, their undesirability, their virtue, or their lack of virtue. Not everyone agrees about what a weed is even though it is a common word that all understand. A few definitions may help illustrate this point.

> "A plant out of place or growing where it is not desired." (Blatchley 1912)

> "A plant whose virtues have not been discovered." (Emerson 1876)

> "Any plant other than the crop." (Brenchley 1920)

> "A plant not wanted and therefore to be destroyed." (Bailey 1941)

> "Those plants with harmful or objectionable habits or characteristics which grow where they are not wanted, usually in places where it is desired something else should grow." (Muenscher 1960)

> "A very unsightly plant with wild growth, often found in land that has been cultivated." (Thomas 1956)

> "Weeds are pioneers of secondary succession, of which the arable field is a special case." (Bunting 1960)

> "A plant is a weed if, in any specified geographical area, its populations grow entirely or predominantly in situations disturbed by man." (Baker 1965)

> "A weed is a plant that originated in a natural environment and, in response to imposed or natural environments, evolved, and continues to do so, as an interfering associate with our crops and activities." (Aldrich 1984)

> "A herbaceous plant not valued for use or beauty, growing wild and rank, and regarded as cumbering the ground or hindering the growth of superior vegetation." (Little et al. 1973)

These definitions range from the poetic of Emerson to the didactic of Baker, Bunting, and Aldrich to the agronomic and control-oriented of Blatchley, Brenchley, and Muenscher. They can create confusion about a very large class of plants that most consider unwanted and undesirable. Most weeds are unwanted, and one key to understanding the confusion about defining a weed is contained in the definition accepted by the Weed Science Society of America, which states that a weed is "any plant that is objectionable or interferes with the activities or welfare of man" (Beste 1983). Man is the focus and it is we who say that a particular plant at a certain place and time is objectionable or interfering with our activities. It does not matter that in another place or time it could be desirable, for now it is a weed and we usually want it out of our environment. We want to control it.

> I will go root away
> The noisome weeds, that without profit suck
> The soil's fertility from wholesome flowers
> *Shakespeare* (1597)

The definition accepted by weed scientists is human, not botanical or ecological, and such definitions often are imprecise and lead to disagreements.

The ultimate etymology of the word "weed" is not known and has been explored by King (1966). Others have given us further evidence of the disagreement about the nature of this large group.

> Once in a golden hour,
> I cast to earth a seed.
> Upon there came a flower,
> The people said, a weed.
> *Tennyson* (1878)

Though a weed is no more than a flower
in disguise,
Which is seen through at once,
If love give a man eyes.

Lowell (1890)

ABBREVIATIONS

The abbreviations used in this book are listed here for convenience of the reader.

A = Arabic
C = Celtic
F = French
G = Greek
Ga = Gaelic
H = Hebrew
Hi = Hindi
L = Latin
M = Malay
S = Spanish
San = Sanskrit
P= Portuguese
Per = Persian

comb. form = combining form
dim. = diminutive
fr. = from
pl. = plural
prob. = probably
var. = variation

Etymology of the Scientific Names of WEEDS

THERE are at least 200,000 species of flowering plants in the world and about 80,000 produce some part that is edible. Humans depend on about 20 of the 50 or so actively cultivated species for food; only 12 provide over 90% of our food calories. Just two grass crops, wheat and rice, occupy over one-third of the world's cultivated land and have fed most of the world's people throughout recorded history. Humans obtain more than one-half their calories from three grasses: rice, wheat, and corn. Ten other crops—soybeans, barley, oats, peanuts, potatoes, millet, sorghum, sweet potatoes, sugarcane, and coconut—provide all but about 5% of the rest of our calories. About 250 plants have been identified as important weeds, which include the large group of weedy species that interfere with growth of crops and the smaller group that interferes more directly with people; for example, poison ivy (*Rhus radicans* L.).

It is surprising that so few families dominate the general category of weeds. Two families, the Poaceae (the grasses) and the Asteraceae (sometimes called the sunflower or composite family), account for about 37% of the world's worst weeds and another 6% are members of the Cyperaceae (the sedge family). Of the world's worst weeds, 68% come from only 12 plant families.

The dominant weed family is the Poaceae, which also includes many of the world's most important crops such as rice, wheat, and corn. Rice feeds more people than any other crop and wheat is grown on more land than any other crop. Other important crops in the Poaceae family include barley, millet, sorghum, oats, and sugarcane. Each major food crop family is also a major weed family; that is, one or more of the world's worst weeds comes from the family. Although it has not been explored in detail, this probably means that our crops and weeds are closely related. It also means that we should look more closely at the largely scorned class we call weeds to find additional food sources. They may be blessings in disguise.

***Abutilon theophrasti* Medic.** Velvetleaf

(Ah-boo′-tih-lon theo-fras′-t)

ABUTILON: (A) a mallowlike plant; also a name for mulberry
THEOPHRASTI: (G) after Theophrastus

This name is a unique combination. *Abutilon,* an Arabic word, means a mallowlike plant and was probably coined by the Arabic philosopher Avicenna, or Ibn-Sina, who lived around 900 B.C. The Arabic name also was a name for mulberry, which has leaves that resemble some species of the genus *Abutilon.* The species name *theophrasti* honors Theophrastus, a Greek philosopher, botanist, and author of the late 4th and early 3d centuries B.C. (372–287 B.C.). He was an early associate of Aristotle and was most likely his successor on the island of Lesbos. His botanical works were probably begun during the middle and last periods of Aristotle's life (died 322 B.C.), and he is regarded as the father of modern Botany.

***Aegilops cylindrica* Host** Jointed Goatgrass

(A′-jil-ops sih-lin′-drih-kuh)

AIGOS: (G) goat; an herb eaten by goats; a kind of wild wheat
OPS: (G) eye
AIGILOPS: (G) a disease of the eyes to which goats are susceptible
KYLINDROS: (G) cylinder; to roll

The specific name is more related to the plant's appearance than is the generic name. The mature seed head looks tightly rolled and is cylindrical. The common name is also helpful because the rachis is divided or joined in segments that easily break off together with the attached seed at maturity. The relationship of the plant to goats or diseases of the eye is obscure, but the fact that the name also means a kind of wild wheat is logical. The Greek etymology calls it a weed of barley, which it is, and it may be eaten by goats.

***Ageratum conyzoides* L.** Tropic Ageratum

(A-ger-a′-tum con-ih-zoy′-deez)

A: (G) not
GERAS: (G) old age
GERONTOS: (G) an old man
AGERATUS: (G) ageless

KONYZA: (G) a strong smelling plant
OIDES: (G) a suffix meaning like or resembling

The generic name is derived from the fact that the flowers do not fade or wither with age. The specific name refers to the plant's strong smell, which resembles some species of the genus *Conyza*.

***Agropyron repens* (L.)Beauv.** Quackgrass

(Ag-ro′-pie-ron ree′-pens)

AGROS: (G) field
PYROS: (G) wheat
REPENS: (L) creeping
REPO: (San) related to
SERPENS: (L) to creep or crawl

This is one of the most common perennial grass weeds in the temperate zones of the world. It spreads by an extensive underground rhizome system and resembles a creeping wheat. The seed and seed head do not exactly resemble those of wheat, but it is close enough for the name to have real meaning.

***Agrostemma githago* L.** Corn Cockle

(Ag-ro′-stem-uh gih′-tha-go)

AGROS: (G) field
STEMMA: (L) wreath or garland, a crown
GITH: (L) a plant of the genus *Nigella;* also a plant with black, aromatic seeds, corn cockle
AGO: (L) a suffix indicating resemblance or connection

We might assume that Linnaeus saw the white hairy stems and the contrasting dark purple to red five-petalled flowers growing abundantly in fields and thought they resembled a garland or wreath. More simply, this common weed of corn fields could have been regarded as the "crown" of the field because of its ornamental qualities. There is no obvious relationship to the genus *Nigella* in that this plant does not have dissected leaves or blue or white flowers. *Gith* was used by Pliny, and the etymology is helpful in the identification of the plant because it does have black seeds.

Allium vineale L. Wild Garlic
(Al′-lee-um vin-e-ah′-lee)

ALLIUM: (L) garlic
VINEA: (L) wine colored, red; of the vineyard
VINEALIS: (L) belonging to vines

Allium is the Latin name for onion or garlic. The specific name may tell us that Linnaeus thought that the pink or red basal portions of the plant or the flowers resembled the red color of wine. He may also have found the weed to be common in vineyards. It is not a viney plant.

Alopecurus myosuroides Huds. Slender Foxtail
(Al′-o-p-cure-us myo-sir-oy′-deez)

ALOPEKOURUS: (G) beardgrass
ALOPEKOS: (G) fox
OURA: (G) the tail
MYOXOS: (G) dormouse
MUS: (G) mouse
OIDES: (G) a suffix meaning like or resembling

The generic name comes from the tail-shaped spike or flowering seed head. The specific name suggests that the spikes resemble a mouse's tail more than any other tail. The author asks the viewer to use some imagination, but the name is not totally devoid of a relationship to appearance.

Alternanthera sessilis (L.)R.Br. ex DC. Sessile Joyweed
(All-ter-nan-ther′-ah sess′-ih-lis)

ALTERNO: (L) to change
ALTERNUS: (L) changing, alternating
ANTHOS: (G) a flower
SESSILIS: (L) sitting; without a stalk

The anthers of some members of this genus are alternately barren. The silver-white flowers occur on compressed spikes in the leaf axils and appear to have no stalk.

Amaranthus hybridus L. Smooth Pigweed
(Am-ah-ran′-thus hi-brid′-us)

AMARANTOS: (G) unfading
MARANTOS: (G) withering
A: (L) not
ANTHOS: (G) flower
HYBRIDUS: (L) a hybrid; a mongrel

Amaranthus retroflexus L. Redroot Pigweed
(Am-ah-ran′-thus reh-tro-flex′-us)

AMARANTHUS: see above
RETRO: (L) back, backward
FLEX: (L) bend

The hybrid *Amaranthus hybridus* is derived from other species. The generic name is the more interesting part of the scientific name. *Marantos* means to fade or wither and the prefix *a-* changes the meaning to not fading or perhaps everlasting. The name is descriptive because the dry calyx and perianth bracts do not fade; they retain their color well past maturity.

One might assume that in *A. retroflexus* the unfading trait refers to the persistent red color of the root. However, this is a specific and not a generic trait as the unfading perianth parts are, so the root is probably not the descriptor Linnaeus had in mind when he chose the generic name. The mature flower head does have a characteristic backward bend, or is recurved.

Ambrosia artemisiifolia L. Common Ragweed
(Am-bro′-z-ah r-tem-iss-ih-fo′-lee-ah)

AMBROSIA: (G) immortality; food for the Gods
AMBROTOS: (G) immortal, divine
ARTEMIS: (G) Diana, daughter of Leto and sister of Apollo
FOLIA: (L) pl. of folium, leaf; akin to many layers of leaves

Ambrosia tomentosa Nutt. Skeletonleaf Bursage
(Am-bro′-z-ah toe-men-toe′-sah)

AMBROSIA: see above
TOMENTUM: (L) stuffing for cushions

Ambrosia trifida L. Giant Ragweed

(Am-bro'-z-ah trif'-ih-da)

AMBROSIA: see above
TRI: (L) three
FINDERE: (L) to divide

Ambrosia artemisiifolia may be a good example of the lack of utility of etymological understanding. *Ambrosia* is Greek and implies immortality and food for the Gods. Dioscorides and Pliny applied the term to one or more herbs and it is now used to refer to herbs with a fragrant, often pungent, although not necessarily attractive, smell. Most of us would not choose this plant as food or as an offering to the Gods. The specific name suggests an association with forests and hills, which is not unreasonable because although the weed is common in agronomic fields, it also is a common weed of open fields. Tying the name to Diana, the virgin huntress and the goddess of forests and hills, is remote from our experience. However, the last part of the specific name does relate to the plant's appearance; it has many layers of leaves.

Ambrosia tomentosa is descriptive for this plant whose leaves are densely covered with soft, downy hairs.

A. trifida, giant ragweed, is usually larger (1–6 m) than common ragweed (0.3–2.5 m) in a given environment, but its specific name arises from large, deeply three-lobed leaves.

Anagallis arvensis L. Scarlet Pimpernel

(An'-ah-gal-liss r-ven'-sis)

ANAGALLO: (G) to decorate
ANA: (G) again
AGALLEIA: (G) to delight in
ANAGALLIS: (G) a plant, probably pimpernel or a chickweed
ARVENS: (L) of the field
ARVUM: (L) arable field

Linnaeus derived the name from the Greek *agalleia* and *ana,* which together mean to delight in or perhaps to laugh and which relate to a fable ascribing to the plant the power to produce mirth and alleviate melancholy. The name may also have been derived from the Greek *anagallo,* to decorate, because of its color and visibility in any field. The five-petalled corolla ranges from scarlet to red, purple, pink, and sometimes blue or white.

Anoda cristata (L.)Schlecht. Spurred Anoda

(An′-o-da cris′-ta-ta)

ANODOS: (G) a way up
ANA: (G) up
HODOA: (G) road, way
CRISTA: (L) tuft, crest
ATUS: (L) suffix meaning possessing

This is an example of a difficult etymology. The name *Anoda* actually came from a Ceylonese colloquial name for *Abutilon,* which was adopted by Cavanilles for this genus of the Malvaceae. The road or way up could refer to the distinct, central purple vein of the mature leaf. The origin of the tuft or crest is also not obvious. Seed capsules are star shaped like a spur and the 15–20 carpels are conspicuously beaked or spurred. There is a modified, spurlike stipule in the leaf axis, which may also be the source of the specific name.

Anthemis cotula L. Mayweed Chamomile

(An′-theme-iss cot′-u-la)

ANTHOS: (G) flower
KOTYLE: (G) cup, cup shaped; anything hollow

Anthemis is the Greek word for the chamomile, and it also shows a relationship to *anthos,* the Greek word for flower. It is not possible to draw a more specific reason for the name unless it relates to the abundant flowers borne by this weed. The specific name *cotula* is not much more meaningful. No specific reference to the plant's appearance is obvious unless Linnaeus saw a resemblance between the central yellow disk flowers of the complex flower head and an inverted cup.

Arctium minus (Hill)Bernh. Common Burdock

(Arc′-t-um mi′-nus)

ARKTOS: (G) a bear
MINUS: (L) smaller, lesser

The rough floral involucre reminds one of the rough hairy coat of a bear. The specific name comes from the fact that this species is smaller than *Arctium lappa* L., great burdock. *Lappa* is from the Latin for burr and also was the name of the root of *A. lappa,* which was used as a diuretic and alterative.

Argemone mexicana L. Mexican Pricklepoppy

(R-gem′-o-nee mex′-ih-ca-nah)

ARGEMONE: (G) a kind of poppy
ARGEMON: (G) small white spot or ulcer on the cornea
MEXICANA: (L) fr. Mexico

Argemone polyanthemos (Fedde)G.B.Ownbey Annual Pricklepoppy

(R-gem′-o-nee poly-an′-theh-mos)

ARGEMONE: see above
POLY: (L) many
ANTHOS: (L) a flower

These annuals have smooth leaves with prickly margins, a blue-green color and yellow or orange-yellow latex. *A. mexicana* has yellow flowers and blue-green foliage with prominent white veins. *A. polyanthemos* is similar but has white flowers and is rarely a biennial. Both are toxic to livestock. The etymology is appropriate because at one time some member of the genus was probably used to cure white specks of the eye. The first species, which originated in Mexico, is related to the second, which, as its name implies, has many flowers.

Artemisia tridentata Nutt. Big Sagebrush

(R-teh-me′-z-ah tri-den-ta′-ta)

ARTEMIS: (G) Diana, the daughter of Leto and sister of Apollo
ARTEMISIA: (G) prob. named in honor of Artemisia of Caria, a noted woman botanist
TRI: (L) three
DENTATUS: (L) toothed
DENTATE: (G) pointed
TRIDENT: (G) three-pointed spear

Diana, daughter of Leto and sister of Apollo, was the virgin huntress, goddess of forests and hills, guardian of women's interests, and was associated with the moon. She was also the Greek goddess of chastity and the one who attended to women's diseases. Thus, this name could have been derived from an understanding of the plant's presumed medicinal value. The medicinal value is further revealed by the common name, sage, which comes from the Latin *salvus,* or safe. The genus could have been named in honor of Artemisia of Caria, a noted woman botanist, medical researcher, and scholar who lived around 400 B.C. in present-day Turkey. Linnaeus accepted the historical name, which he said came from St. Johannis. The leaves of this common weed of rangeland in the arid western United States have three blunt tips (tridentate), and the name is a good description of the appearance of the leaf tip.

Asclepias speciosa **Torr.** Showy Milkweed

(Ass-cle′-p-ass spee-c-o′-sa)

ASKLEPIOS: (G) god of medicine, son of Apollo and Coronis, physician hero of Greek medicine
SPECIOSUS: (G) beautiful, showy

Asklepios, Greek god of medicine and healing, was the son of Apollo and Coronis and a physician hero of Greek medicine. The name actually honors the Greek herbalist Aesculapius—chief among physicians. He was a supposedly historical person who used the knowledge and power of herbs taught to him by the centaurs to become a physician. The story is convincing to the modern mind until we read the part about centaurs. These characters of Greek mythology were a race of monsters born of Ixiom having the head, arms, and trunk of a man and the body and legs of a horse. However, a little bit more investigation reveals that the Centaurs were also a primitive Thessalian tribe. It could all be true. The plants of this genus have medicinal value. The milky latex was used to enhance passage of body fluids and to increase human milk flow. Later it was found to contain a digitalislike compound, which gave a physiological basis for its prior use. The specific name is appropriate because the flowers are showy and attractive parts of the landscape unless they inhabit your crop field.

Astragalus bisulcatus **(Hook.)Gray** Twogrooved Milkvetch

(Ass-tray′-gah-lus bi-sul-k′-tus)

ASTRAGALUS: (G) ancient name of a legume plant, milk-vetch; also a bone of the human vertebra
BI: (L) two
SULCATUA: (L) furrowed
SULCUS: (L) furrow or groove

Astragalos is the name of one of the bones of the human spinal column and alludes to the seed shape of members of this genus. The specific name refers to two grooves on the upper side of the seed pods. Many species of this large genus are poisonous.

***Atriplex canescens* (Pursh)Nutt.** Fourwing Saltbush

(At′-trih-plex can′-s-enz)

ATRIO: (L) *atrium* = cavity or chamber
PLEX: (L) comb. form meaning to fold
PLICARE: (L) to fold
ATRIPLEX: (L) a saltbush
CANESCERE: (L) to be gray or white
CANESCO: (L) to become white or hoary

***Atriplex patula* L.** Spreading Orach

(At′-trih-plex pat′-u-la)

ATRIPLEX: see above
PATERE: (L) to be open
PATULUS: (L) spread out, open

The pistillate (female) flowers of *Atriplex* plants are subtended by a pair of triangular bracts. When enlarged, these bracts surround the fruit to form a utricle, a small, one-seeded fruit in a bladder. The bladder or utricle is the atrium or chamber that is plexed or folded. The species *A. canescens* has gray-white or hoary foliage.

The species *A. patula* was so named because the sessile fruiting bracts of the perianth are united only at their rounded base and are thus largely open.

***Avena fatua* L.** Wild Oats

(A′-v-na fat′-u-ah)

AVENA: (L) oats
FATUUS: (L) silly, foolish

With the specific name of *fatua,* meaning silly or foolish, weed scientists would readily agree with this description of the wild oat. Anyone who allows it to grow can be regarded as silly or foolish because it depresses crop yield, especially of wheat and barley, in which it appears with disturbing regularity. It is not valued for food or feed and farmers are silly to have it and foolish to let yield be depressed by it. However, from nature's point of view the plant is not at all silly or foolish because it so successfully competes with crops we plant. The name illustrates how plants are named to suit perceptions of their value and not necessarily to fit how they perform in the natural world. Weediness, and its associated undesirability, are in the eyes of the beholder and those who name the plants.

Axonopus compressus **(Sw.)Beauv.** Tropical Carpetgrass

(Ax-o-no′-pus com-press′-us)

AXON: (G) axis
POUS: (G) foot
COM: (L) *cum* = with
PREMERE: (L) to press

This plant seems appropriately named. It is a creeping, perennial, mat-forming grass with blunt, rounded leaf tips. The flowering culms arise from a single axis, or foot, and are one-flowered spikelets on a one-sided spike-like raceme. The flowering structure is laterally compressed and the leaf sheath is similarly strongly, laterally compressed.

Barbarea vulgaris **R.Br.** Yellow Rocket

(Bar-bar′-e-ah vul-gare′-iss)

BARBAREA: (L) after St. Barbara
EA: (L) ending of generic name taken from personal name
VULGARE: (L) to make known
VULGATUS: (L) ordinary, common
VULGUS: (L) mob, common people

St. Barbara was a 3d century B.C. Christian martyred by her pagan father for her conversion to Christianity. Her father gave her to Martian, the governor of Nicomedia, who tortured her. As her father was about to cut off her head, he was struck dead by lightning and died at her feet. St. Barbara is invoked against lightning and is known as the patron saint of arsenals and powder magazines. Tradition tells us that seeds of one member of the genus were sown around the mid-December feast of St. Barbara in Europe. Any relation to the plant's looks or behavior is not obvious. *Vulgaris* means common and is associated with the names of many plants that were regarded as common or ordinary because they occurred so frequently. Many weeds fall into this category.

Bidens pilosa **L.** Hairy Beggarticks

(B′-dens pie-lo′-sa)

BI: (L) two
DENTI: (L) a tooth
DENTATUS: (L) toothed
PILOSUS: (L) hairy, usually short, soft hairs

The etymology of this name is helpful when trying to identify mature plants. The angled achenes (seeds) are commonly two-awned but are sometimes four-awned. The two-awned achene is the origin of the name for it gives the appearance of being bidentate or two-toothed. The specific name is not quite as obvious. The plant is not especially hairy and it definitely does not have abundant, soft, short hairs as the name leads one to expect. The leaf margins are hairy but not so much that the observer would call the plant hairy. It is more likely that the name was derived from the fact that the achenes with their two-barbed awns are good at lodging in the hair of animals, an excellent technique for achieving distribution.

Borreria laevis **(Lam.)Griseb** Buttonplant

(Bor-rare′-e-ah lay′-viss)

BORRERIA: after William Borrer, a British botanist, died 1862
LAEVIS: (L) *lavis* = having a smooth, polished surface; shining, without hair

Many genera have been named after distinguished botanists. The specific name reveals that some portion of the plant is smooth and shining. In this case it is not the stem, which does have some hairs. It is more likely that the word refers to the author's perception that the entire plant, especially the leaves, which appear smooth, is without hairs.

Brachiaria mutica **(Forsk.)Stapf.** Paragrass

(Brack-e-air′-e-ah mew′-tih-ca)

BRACHYS: (L) short
ARIA: (L) like or connected with
BRACHIUM: (L) short arm
MUTICUS: (L) curtailed, cut off

The annual and perennial grasses in this genus are all accurately described by the generic name, which refers to the paired, 2.5- to 3.5-mm-long spikelets borne on racemose branches of the spreading purple panicle. One of the paired spikelets has a distinct, but short, stalk. In this species spikelets and their associated seeds are awnless, or curtailed.

Brassica kaber (DC.)L.C.Wheeler Wild Mustard
(Brass'-ih-cah kay'-bur)

BRASSICA: (L) cabbage
BRESIC: (C) cabbage
KABER or
CABER: (G) pole; beam

Brassica nigra (L.)W.J.D.Koch Black Mustard
(Brass'-ih-cah nye'-gra)

BRASSICA: see above
NIGER: (L) black, dark

The generic name reveals that the plants are related to cabbage. *Kaber* refers to the strong 50- to 10-cm-tall stem, whereas *niger* refers to the black or red-brown color of mature seeds.

Bromus tectorum L. Downy Bromegrass
(Bro'-mus tek'-tor-um)

BROMOS: (G) grasses, ancient name for the oat
BROMATOS: (G) food
TECTOR: (L) in general, straw for thatching
TEGO: (L) to cover
TECTULUM: (L) roof, covering

This weed is not a good fodder crop because awns on the seed make it difficult, if not unpleasant, for animals to chew. It is part of a large genus with members that are suitable for forage and hay. Perhaps the best example is *Bromus inermis,* a common pasture grass. The small plant could have been used for thatching roofs.

Calopogonium muconoides Desv. Calopo
(Cal-o-po-go'-ne-um mew-co-noy'-deez)

CALOS: (L) beautiful
POGON: (G) beard
IUM: (G) dim. suffix meaning belonging to; dim. suffix
MUCOR: (L) mold, moldiness
MUCERE: (L) to be moldy
OIDES: (G) suffix meaning like or resembling

This trailing, decumbent, perennial vine can twine and climb when offered a chance. The stem bends and changes course frequently at its thickened nodes where purple to blue flowers are clustered. When the French botanist Augustin Nicoise Desvaux (died l856) named the plant, he saw purple to blue flowers with a beautiful lip. The basal petal of the bilabiate corolla is linear-oblong and forms a liplike structure. It has numerous club-shaped hairs above it, and the petal is hairy at its apex. Beards can be beautiful! The specific designation *mucor* is not related to appearance and is probably related to the author's perception of a moldy smell when the fresh plant was crushed.

***Cannabis sativa* L.** Marijuana

(Can′-ah-biss sah′-t-va)

CANNABIS: (L) hemp, origin obscure
KANNABIS: (G) hemp
SATIVA: (L) sown
SATIVUS: (L) that which is sown
SATUS: (L) a planting
SERO: (L) to sow

The ultimate etymology of the generic name is obscure beyond the Latin word for hemp. The plant was, and still could be, used for fiber-making but is now grown primarily for its narcotic properties. Its illegality makes the specific name imprecise. It is planted or sown but it is not desired, except by those who grow or obtain it illegally.

***Capsella bursa-pastoris* (L.)Medic.** Shepherdspurse

(Cap′-sell-ah bur-sa-pas-tor′-iss)

CAPSA: (G) box, case
ELLA: (L) suffix meaning small
BURSA: (L) a hide or skin; purse, usually made of skin
PASTOR: (L) a herdsman, shepherd
PASTORIS: (L) of the shepherd, of the keeper

Shepherds of the middle-eastern part of the world still carry a purse or small case that resembles the seed pod of the many-seeded, two-valved seed pod of this common annual weed. This is one of the best examples of how accurate the fit between a scientific name and a plant's appearance may be. However, with such an obscure origin it is not surprising that the etymology is not used more in teaching.

***Cardaria draba* (L.)Desv.** Hoary Cress
(Car-dare'-e-uh drah'-ba)

KARDIA: (G) heart
DRAB: (G) dull, lifeless, faded
DRABE: (G) sharp, burning; also a kind of cress
DRABA: (G) a name applied to a cress by Dioscorides

This vigorous and rather unattractive perennial has heart-shaped, often inflated, 2- to 3-mm-long seed pods, which are the source of the generic name. The foliage is pale green and can be described as dull and lifeless, although the farmer who has to face this weed will see it as quite lively. It is not eaten by any animals if they have a choice. The foliage has a sharp taste, which could be the source of the specific name.

***Carduus nutans* L.** Musk Thistle
(Car'-dew-us new'-tans)

KARDO: (G) thistle
CHARDON: (F) thistle
CARDE: (F) edible leaf or stalk of the cardoon or artichoke
CARD: (C) combing wool
NUTARE: (L) to nod again and again
NUTO: (L) to nod

It is true that, like many wild plants, portions of this plant can be eaten but it is hard to find anyone who does so regularly. Many of us would be pleased to find how good they taste, but we prefer our food from the supermarket and do not explore tastes of the natural world. This is definitely a spiny thistle, which makes eating it a little difficult, and its spiny nature makes use of its seed heads for carding wool questionable. The specific name comes from the appearance of its large flower heads, which droop and nod.

***Cassia obtusifolia* L.** Sicklepod
(Cass'-e-uh ob-too-sih-fo'-lee-ah)

CASIA: (L) a bark resembling cinnamon
CASSIDIS: (L) a helmet
OBTUSUS: (L) dulled, blunt
FOLIA: (L) leaf

The name *Cassia fistula* was given to one species of the Pudding Pie tree, a native of India, which produced cassia pods containing a pulp used as a laxative. In one derivation *casia* means resembling cinnamon, but this does not apply to all members of the genus. It is more likely that the generic name was assigned because of the hooded nature of the usually yellow corolla; it hangs downward, like a helmet, to cover the rest of the floral structure. The specific name *obtusifolia* applies precisely because the leaves are obtuse, without any points or protrusions.

***Celosia argentea* L.** Celosia

(Cell-o′-c-ah r-gen′-t-uh)

KELOS: (G) dry, parched
ARGENTUM: (L) silver

The generic name is descriptive, for the flowers acquire a seared, parched appearance with age. The specific name is an apt description of early flower color. Both names therefore apply to flowers and the rest of the plant is omitted from the scientific nomenclature. No one has ever made a rule that scientific names have to be completely descriptive.

***Cenchrus echinatus* L.** Southern Sandbur

(Senk′-rus etch-ih-nat′-us)

KENCHROS: (G) millet
ECHINATE: (L) armed with numerous rigid hairs or spines
ECHINUS: (L) sea urchin

***Cenchrus incertus* M.A.Curtis** Field Sandbur

(Senk′-rus in-cert′-us)

CENCHRUS: see above
IN: (L) not
CERTUS: (L) settled, sure

Cenchrus echinatus is a member of the Paniceae tribe of the Poaceae family. Each of the other six genera in the tribe has important weedy representatives. The *Panicum* genus also includes the world's millets. Thus, there is a close relationship to millet recorded in the generic name. The specific name is very appropriate when one sees, or better feels, the very spiny seed bur.

Not too long ago *C. incertus* was called *C. pauciflorus,* a milletlike plant with few flowers. Moses Ashley Curtis (died 1872), an American botanist, changed the name to *incertus* perhaps because he was not sure of its precise taxonomic placement or because it is a confusing and variable species.

Centaurea repens **L.** Russian Knapweed

(Sen-tor′-e-ah ree′-pens)

CENTAUR: (G) spearman, piercer
REPEN: (L) creeping

The generic name comes from the classical name of a plant in Grecian fables, which is said to have cured a wound in the foot of Chiron, one of the race of Centaurs, a primitive tribe of Thessaly, the wild horse breakers. They were also characters in Greek mythology, born of Ixiom and having the upper torso of a man and the body and legs of a horse. The specific name comes from the fact that the plant is a perennial with strong, black, creeping roots that can send up new shoots from frequent nodes.

Cerastium vulgatum **L.** Mouseear Chickweed

(Sir-ass′-t-um vul-gay′-tum)

CERAST: (L) horned
KERAS: (G) a horn; a bow
IUM: (G) dim. suffix
VULGATUS: (L) ordinary, common
VULGARE: (L) to make known
VULGUS: (L) mob, common people

The seed capsules of this annual weed appear like horns as they emerge from the calyx, and they retain their horn shape when mature. Variations on the Latin *vulgare* are common specific names for plants that are indeed common or ordinary. One might say that some weeds are so common as to be considered ubiquitous, and they are found almost everywhere that agriculture is practiced. However, the name could also be derived from the fact that weeds, like this one, commonly occur around dwellings, are readily observed, and thus are considered common.

Ceratophyllum demersum L. Coontail

(Sir-at-o-fill′-um d-merse′-um)

CERAST: (L) horned
KERAS: (G) horn; a bow
PHYLLON: (G) leaf
DEMERSUS: (L) submerged

This perennial, rootless, aquatic weed has small leaves that look like deer antlers. The threadlike leaves, arranged in whorls around the stem, are mostly two-forked with two to five (mostly four) segments with scattered, minute, spiny teeth. The specific name is derived from its completely submerged growth.

Chenopodium album L. Common Lambsquarters

(Key-no-po′-d-um al′-bum)

CHEN: (G) a goose
PODOS: (G) a foot
PODION: (G) a small foot
ALBUM: (L) white

The leaves of this common annual weed and its relatives in this genus resemble the foot of a goose that has white feet. Linnaeus must have seen the resemblance when he chose the name. The plant looks green when viewed from a distance, but the underside of the leaves has a gray-white mealy covering. It is this and perhaps the sometimes pale green color of the whole plant that make the specific name appropriate.

Chromolaena odorata (L.)R.M.King & M.Robinson Bitterbush

(Crome-o-lay′-na o-dor-ah′-ta)

CHROMA: (G) colored
LAENA: (G) a cloak
ODORUS: (L) fragrant
ODORATUS: (L) sweet smelling

Young seedling leaves range from green to dark purple; older leaves are green but the flowers are the true source of color. The inflorescence is in axillary or terminal clusters of 10–35 light blue to white flowers. One way to begin to identify the plant, at any growth stage, is to crush a few leaves and smell them. They emit a powerful, pungent smell, and this yields the specific name.

Chrysopogon aciculatus (Retz.)Trin. Pilipiliula

(Cry-so-po'-gon a-sic-u-lay'-tus)

CHRYSOS: (G) gold
CHRYSEOS: (G) golden
POGON: (G) beard
ACICULA: (L) a point

This weed does not quite have a golden beard; in fact, it may take some imagination to see any beard at all. The inflorescence of this creeping perennial is a 3- to 10-cm-long narrow panicle with several whorls of short, slender, horizontal, red-tinged branches. It is sort of a weak, perhaps a teenaged, beard. Two fine, sharp bristles on the caryopsis give the specific name.

Chrysothamnus nauseosus (Pallas)Britt. Gray Rabbitbrush

(Cry-so-thame'-nus naw-z-o'-sus)

CHRYSOS: (G) gold
CHRYSEOS: (G) golden
THAMNOS: (G) a shrub
NAUS: (G) a ship
NAUSIA: (G) nausea, sea-sickness
OSUS: (L) a suffix meaning full of

This common rangeland weed in the western United States blooms in a mass of golden flowers; the entire plant looks like a large golden shrub. It is also nauseating to cattle, which do not eat it unless starved. It combines visible beauty with invisible digestive upset and the name fits in both respects.

Cichorium intybus L. Chicory

(Chih-cor'-e-um in-tie'-bus)

KICHORA: (G) chicory; prob. fr. Arabic
INTYBUS: (L) endive, chicory

The generic name is Greek and was probably derived from Arabic, but we do not know its Arabic meaning. The specific name means endive or succory, but further elucidation of its meaning was not possible.

Cirsium arvense (L.)Scop. Canada Thistle
(Sear′-c-um r′-vence)

KIRSION: (G) a thistle
KIRSOS: (G) a swollen vein
ARVENS: (L) of the field
ARVUM: (L) arable field

Cirsium vulgare (Savi)Tenore Bull Thistle
(Sear′-c-um vul′-gare)

CIRSIUM: see above
VULGATUS: (L) ordinary, common
VULGARE: (L) to make known
VULGUS: (L) mob, common people

The meaning of this generic name has little, if any, relationship to the plant's weedy character. The name comes from the Greek *kirsos,* which literally means a swollen vein. It is derived from a former use of thistles (not necessarily this one) to treat swollen veins. One can imagine that a poultice was made from the plant and applied to an infected area. Now this perennial, spiny, very persistent plant is almost universally regarded as weedy, whether it occurs in a lawn or an arable field. The specific name *arvense* is apt since the plant is rarely cultivated, but common in arable or cultivated fields.

The specific name *vulgare* implies that it is more common than *C. arvense.* This may have been true when they were named. Perhaps it is related to the way we now practice agriculture, but the perennial *C. arvense* is the more common of the two.

Commelina benghalensis L. Tropical Spiderwort
(Co-meh-line′-uh beng-ha-len′-sis)

COMMELINA: after Kaspar Commelin, a Dutch botanist, died 1731
BENGHAL: (L) fr. Bengal, India
ENSIS: (L) suffix meaning place of origin

Commelina diffusa Burm.f. Spreading Dayflower
(Co-meh-line′-uh dih′-few-sa)

COMMELINA: see above
DIFFUSA: (L) spread out, diffuse

The genus was named after two Dutch botanists who worked in the early 17th century. The humor of Linnaeus shows in the name that was picked because flowers of *Commelina* have three petals: two showy, while the third is inconspicuous. The two botanists were Jan and Kaspar Commelin but a third, also named Commelin, died before accomplishing anything of lasting significance in Botany. The first specific name, *benghalensis,* indicates that the plant is native to, or was first identified in, the state of Bengal in eastern India. The second specific name, *diffusa,* indicates that this plant may be more widely dispersed or more spread out than *C. benghalensis.* Both plants are prostrate, sprawling, succulent, and annual or perennial herbs with ascending flower stalks. The name is probably related to its environmental dispersal rather than its habit of growth.

Convolvulus arvensis **L.** Field Bindweed

(Con-vol′-view-lus r-ven′-sis)

CONVOLVO: (L) to roll together, roll up
CON: (L) together
VOLVERE: (L) to roll
ARVENS: (L) of the field
ARVUM: (L) arable field

The common and scientific names both fit very well. This vigorous, creeping perennial is capable of spreading aboveground by its long trailing stems and underground by its extensive creeping root and rhizome system. The stems can actually bind a whole field together, and there is only a little hyperbole in the story of the farmer with such a bad bindweed problem that a stalk on one corner of the corn field could be used to shake the entire field.

Conyza canadensis **(L.)Cronq.** Horseweed

(Con′-eez-uh can-uh-den′-sis)

KONYZA: (G) a strong smelling plant
CANAD: (L) fr. Canada
ENSIS: (L) suffix meaning place of origin

Plants with the generic name *Conyza* usually have a strong smell, much like the smell of turpentine. Horseweed is very common but lacks a distinctive personality. It is so common that it is often unnoticed. It has never established itself as a major weed problem and exists around the fringes of cropped fields, in pastures, and neglected areas. A frequent common name is fleabane because its strong smell is supposed to be obnoxious to fleas. Its origin is Canada, where it is very common.

***Corchorus olitorius* L.** Nalta Jute

(Cor′-cor-us ahl-ih-tor′-e-us)

KORCHOROS: (G) a plant of bitter taste
KORIS: (G) a kind of St. Johnswort
CORCORUS: (L) a poor kind of pulse (legume)
OLITOR: (L) a kitchen garden
OLITORIUS: (L) of or pertaining to vegetables

Corchorus olitorius has a bitter taste, as many other plants do. It may have been a cultivated herb used to season vegetables but was probably not used as a vegetable.

***Cuscuta campestris* Yuncker** Field Dodder

(Cuss′-q-tuh cam-pes′-tris)

CUSCUTA: (L) dodder
(H) to bend
(A) *chasuth, keshut* = the name for the plant
CAMPESTRIS: (L) pertaining to fields

Cuscuta, meaning to dodder or bend, could have come from the Arabic name for the plant—chasuth or keshut. It emerges from soil as a colorless stem, rotates to seek a host, and continues its life as an obligate stem parasite. If it does not find a suitable host, it dies. The dictionary reveals that to dodder is to tremble or shake from weakness or old age; we speak of doddering old fools. It also means to progress feebly. The adjective doddered means deprived of branches. When *Cuscuta* first emerges it is weak and it trembles in the slightest breeze as it searches for its host. A mature plant consists of slender, threadlike, mostly yellow stems, which twine and coil about host plants. The entire stand can then be seen to tremble with the slightest breeze. It is a serious weed of many crops and can eliminate the ability to grow the crops it parasitizes.

Cynodon dactylon (L.)Pers. Bermudagrass
(Sin′-o-don dak′-til-on)

KYONOS: (G) a dog
ODONTOS: (G) tooth
DAKTYLOS: (G) finger, toe

The generic name means a dog's tooth and is probably related to the leaf on rootstocks and runners; these leaves are short, thick, and scalelike. If one has a good imagination they may even appear to look like a dog's tooth. It is a highly prized perennial pasture grass in some warm environments, but in others it is despised as a very competitive weed. It creeps aboveground by stolons and underground by rhizomes, rooting at the nodes and easily forming dense mats. The specific name is more obviously related to the plant's appearance. The erect, flower-bearing stem grows 10–50 cm high. Its flowering spikes are about 4 cm long with three to five in a cluster. These spikes radiate out from the culm like fingers.

Cyperus difformis L. Small-flower Umbrella Sedge
(Sigh′-per-us dih-for′-mis)

KYPEIROS: (G) rush, sedge
DIFFORMIS: (L) irregular, unevenly formed

Cyperus esculentus L. Yellow Nutsedge
(Sigh′-per-us s-cue-len′-tus)

CYPERUS: see above
ESCULENT: (L) edible, good to eat
ULENTUS: (L) a suffix meaning abundance
ESCA: (L) food

Cyperus iria L. Rice Flatsedge
(Sigh′-per-us ear′-e-ah)

CYPERUS: see above
IRIDOS: (G) the rainbow; the iris of the eye
IRIONIS: (L) a kind of cress
IREOS: (L) obscure; prob.fr. an altered form of iris

Cyperus rotundus L. Purple Nutsedge
(Sigh′-per-us ro-tun′-dus)

CYPERUS: see above
ROTUNDO: (L) circular, round

The generic family name Cyperaceae comes from the Greek word for rush or sedge. These plants are not grasses although they are often confused with grasses. A distinguishing trait is the triangular cross-section of the stem, with leaves in three rows or ranks. Grasses have cylindrical, hollow stems with leaves in two rows. The specific name *difformis* is undoubtedly related to floral structure. The inflorescence is a dense, globose, simple, or compound umbel. Primary rays of the umbels are 2–4 cm long and secondary rays are about 1 cm long, which already depicts an irregular structure. In addition, secondary rays may be sessile or have a long peduncle, another irregularity. These irregularities are, of course, our perception of structure. Given its role in nature and its ability to survive under the acute stress of crop competition, the supposed irregularities may become mechanisms of survival and advantages in the competitive struggle that often characterizes a cropped field.

Cyperus esculentus was named for its edible tubers, which are anatomically similar to potato tubers (*Solanum tuberosum* L.) but are often called nuts, giving the common name nutsedge. However, collecting the tubers for eating is rare. They are 1–2 cm long, commonly globose, and slightly flattened transversely. The surface is smooth because leaf scales, present early, disappear with maturity. The taste is similar to an almond but somewhat sweeter with what has been called an earthy flavor. It used to be possible to find nutsedge tubers in markets in the southern United States under the name chufa. *C. esculentus* foliage has a yellow cast.

C. iria has an origin related to the eye in an obscure way. One source says that *ireos* is an altered form of iris, which was arbitrarily applied to white-flowered species in contrast to purple-flowered ones. We do not know why this was done. The problem arises when we discover that the flower spikelets give an overall appearance of being yellow-brown to green and are not at all white.

C. rotundus is called purple nutsedge because of the red-brown to purple color of mature, overlapping florets. Its tubers are covered with a persistent red-brown coat, but this probably did not give rise to the common name. The plant is generally round in cross-section when viewed from above; this is probably the origin of the specific name.

***Dactyloctenium aegyptium* (L.)Willd.** Crowfootgrass

(Dak-tih-lock-ten′-e-um e-gyp′-t-um)

DACTYLO: (L) comb. form
DACTYLOS: (G) finger, toe
KTENION: (G) a little comb
TENER: (L) soft, delicate

AEGYPT: (L) Egypt
IUM: (L) dim. suffix meaning belonging to

Dactyloctenium is difficult to pronounce but easy to define, and the name fits this weedy species. The inflorescence is an erect, 15- to 60-cm-long stalk topped by two to five fingerlike, digitate or pectinate terminal spikes. Each of these digits or spikes is 2–5 cm long and 5–8 mm wide. Small spikelets are crowded on the main spike. They are not especially soft and the spike terminates in a sharp point. Spikes can be regarded as tender or delicate because they shatter easily when mature. The mature spike resembles a small comb. *D. aegyptium* apparently is a native of Egypt but is now a widespread tropical weed.

***Datura stramonium* L.** Jimsonweed

(Dah′-tur-uh stra-mo′-nee-um)

DHATURA: (H) a genus of poisonous plants; the thornapple
TATORAH: (A) the thornapple
STRAMONIUM: (L) perhaps an alteration of the Tartar word *Turman* = medicine for horses
STRYCHNOS: (G) nightshade
MANIKOS: (G) mad

As stated in the Introduction, this plant has a very descriptive common name. It tells us nothing about what the plant looks like or how it behaves, but it is a good story with a valid historical origin. As shown above the generic and specific names also do not tell us anything about how the plant looks. But Linnaeus chose to retain what he called the barbarous (i.e., not derived from Latin or Greek) name *Datura* because he could relate it to the Latin *dare,* to give; it was given to those whose sexual powers were weak or enfeebled. The specific name alludes to past uses. Of the plant names reported herein, this is the only one with an origin from Hindi, one of the major languages of India. However, even that is not sure because of possible Arabic and Greek origins. It was not possible to determine which was the older usage.

***Daucus carota* L.** Wild Carrot

(Daw′-cus care-ott′-ah)

DAUKOS: (G) carrot
CAROT: (G) stupor
(L) carrot
CAROTA: (C) red color

The Greek generic name means carrot and aptly describes the plant because the root of this common weed resembles a garden carrot, *Daucus sativa* L., except it is smaller in diameter and white. It is also slower growing and usually occurs in areas with low fertility. The garden carrot produces its rich, orange-colored root in one season whereas the wild carrot, a biennial, produces a taproot and only a flat rosette of leaves during the first year of its life. In the second year it flowers by drawing on root reserves. The name may also come from the Greek *daio* (I burn) because of its use as a poultice and "heating" plant in ancient medicine. Others suppose the name to be of Celtic origin (*car* = red), referring to the red color of the root in the domestic carrot but not in the wild weed species. Flowers and leaves give this plant its most frequently used common name, Queen Anne's lace. During the reign of Queen Anne of England (1702–1714) the finely divided compound leaves were inserted in ladies hair and in floral bouquets.

***Delphinium geyeri* Greene** Geyer Larkspur

(Del-fin′-e-um guy′-er-i)

DELPHINION: (G) larkspur
DELPHINOS: (G) dolphin
GEYERI: after Carl A. Geyer, a German botanist, died 1853

All members of this genus are perennial herbs, some of which are poisonous weeds of rangelands in the western United States and other range areas of the world's temperate zones. They have palmately divided leaves, flowers in terminal racemes, and are related to the garden delphinium, *Delphinium ajacis* L. The generic name was given because of the shape of the nectary, or spur. There are five sepals, the upper one being extended into a characteristic spur, which, for some, resembles the shape of a dolphin's body. It is beautifully curved and gives the pretty, frequently blue or purple flowers a distinctive look and the plant its common name, larkspur.

***Descurainia pinnata* (Walt.)Britt.** Tansy Mustard

(Des-cur-ain′-e-ah pin-nah′-tah)

DESCURANIA: after Francois Descourain, a French botanist, died 1740
PINNATUS: (L) feathered, winged

This is another example of a genus named after a botanical colleague, in this case, the French botanist Francois Descourain. The specific name *pinnata* literally means feather-formed and describes the once or twice divided, fine leaf segments.

***Desmodium tortuosum* (Sw.)DC.** Florida Beggarweed

(Dez-mo′-d-um tor-chu-o′-sum)

DESMOS: (G) a bond; like a chain
DESMIDOS: (G) a bundle
IUM: (L) dim. suffix
TORTUS: (L) a twisting
TORTUM: (L) to twist

The plant part most like a chain and the most likely source of the generic name is the fruit. The pod is 5–12 mm long with two to six deeply segmented joints. Each segment contains one seed. Flowers occur in terminal bundles or clusters of pealike blooms. This is an annual or short-lived perennial with short, hooked, or twisted hairs on each seed pod.

***Digitaria ciliaris* (Retz.)Koel.** Southern Crabgrass

(Dih-jih-tare′-e-uh sil-e-r′-iss)

DIGITUS: (L) a finger, toe
ARIA: (L) pertaining to, resembling
CILIOLA: (L) hair or hairlike
ARIS: (L) pertaining to

***Digitaria sanguinalis* (L.)Scop.** Large Crabgrass

(Dih-jih-tare′-e-uh san-guih-nay′-liss)

DIGITARIA: see above
SANGUIS: (L) blood
ALIS: (L) pertaining to

***Digitaria scalarum* (Schweinf.)Chiov.** Blue Couch

(Dih-jih-tare′-e-uh ska-lar′-um)

DIGITARIA: see above
SCALARIS: (L) pertaining to a ladder
SCALA: (L) stairs, ladder

Digitaria comes from the ascending, spreading panicle of digit- or fingerlike racemes, each 5–10 cm long. *D. ciliaris* often has a whorl of four to nine of these at the top of the central stem. Sometimes they are arranged along a short, common axis. Sharp, pointed spikelets, densely crowded in two rows along one side of the raceme, frequently develop a fringe of brownish to red hairs along their margins. These usually do not appear until the plant is mature, but they are the origin of the specific name.

In modern English to be sanguine is to be confident and optimistic, body color is good with a robust appearance, and one seems sturdy and marked by cheerfulness. However, sanguine also means that something consists of or is related to blood. The specific name *sanguinalis* applies to this species because the flowering glumes, lemmas, and palea, or the upper of the two enclosing floral bracts, are gray-brown to purple. The purple or bloodlike color often dominates late in the growing season. It is not infrequent to have the entire plant become purple late in the season especially in lawns. This, of course, gives lawns a varied color through the season, but most people do not regard this as an advantage and want to rid their lawn of this annual weed. However, we might also assume that the name was given because of the sturdy, robust nature of this weed in lawns and gardens.

The flowering inflorescence of *D. scalarum* is more spread out and there is a ladderlike arrangement of spikelets on the raceme.

***Dipsacus fullonum* L.** Common Teasel

(Dip′-sa-cus full-lo′-num)

DIPSA: (G) thirst
DIPSAKOS: (G) a kind of teasel
FULLARE: (L) to full

The generic name refers to a cavity formed by united, cup-shaped, opposite leaf bases, which clasp the stem to form a cavity that traps water. This biennial flowers and sets seed in its second year. Flowers form in conical heads surrounded by long ascending bracts. Receptacles are chaffy with long, tapering, flexible, straight-pointed bristles. This bristly structure has been used by fullers who were people who shrank and thickened wool to make it fuller. The seed heads were used to card wool (disentangle fibers and raise the knap of any fiber) before it was thickened (made full) by moisture, heat, and pressure.

Echinochloa colonum (L.)Link Junglerice
(E-kine-o-clo′-uh co-lo′-num)

ECHIN: (L) prickly
ECHINOS: (L) sea urchin; hedgehog
CHLOA: (G) grass; young herbage
COLO: (L) to inhabit
COLONUS: (L) colonist; farmer
COLONA: (L) a man of the country

Echinochloa crus-galli (L.)Beauv. Barnyardgrass
(E-kine-o-clo′-uh cruise-gal′-e)

ECHINOCHLOA: see above
CHLOA: see above
CRUS: (L) the lower part of the stalk; leg or thigh
GALLI: (L) pl. of gallus-cock

Echinochloa glabrescens Munro ex Hook.f. Barnyardgrass
(E-kine-o-clo′-uh glab′-res-cans)

ECHINOCHLOA: see above
GLABRESCO: (L) to grow smooth bare, bald
GLABER: (L) smooth

If you have encountered a sea urchin or hedgehog, especially when you weren't expecting to, you will have some awareness of the appropriateness of the generic name. The bristling, short awns of the spikelets on the raceme are quite prickly. They are not as sharp as the spines of sea urchins nor as long as the hair of a hedgehog but there is a resemblance, albeit one that may stretch your imagination. *E. colonum* is a weed of the country, especially rice country, and one of the most important rice weeds in the world.

Origin of the specific name *crus-galli,* whose parts mean the lower part of the stalk or leg and a cock, is not obvious, but one can make some educated guesses. The inflorescence is often tinged with pink or purple and may resemble a cock's comb. It has densely crowded spikelets, and the lowest and longest branches often rebranch and spread at maturity. This spreading and coloration may also resemble a cock's comb but the resemblance is superficial. It is more likely that Linnaeus and A. M. F. J. Palisot de Beauvois may have seen a resemblance to the cockspur in the floral awns, which could be the shank or place where the spur grows. They may also have thought the entire floral structure resembled the gallus or cock's comb.

The specific name *glabrescens* means smooth and is related to the botanical term glabrous, which means without hairs. Leaf sheaths of this species lack hairs, whereas other species have some hairs on the leaf sheath.

***Eclipta alba* (L.)Hassk.** (syn. *E. prostrata* L.) Eclipta

(E-clip′-ta al′-ba) (pros-tray′-ta)

EKLIPSIS: (G) abandonment
EKLEIPEIN: (G) to abandon
EKLEIPO: (G) to cease, stop; to be deficient
ALBUS: (L) white
PROSTRATA: (L) prostrate

Eclipta alba is not a forsaken plant although many farmers who have it in their rice crop wish it would abandon them. It was called *Eclipta* because of the absence of a pappus on the small, dry, hard achene of many species. The specific name comes from the white disc florets of the inflorescence. The synonym *E. prostrata* was named because of its prostrate growth habit.

***Eichornia crassipes* (Mart.)Solms** Water Hyacinth

(I-cor′-nee-uh crah-sip′-eez)

EICHORNIA: after Johann A. F. Eichorn, a Prussian minister of education, died 1856
CRASSUS: (L) thick; heavy
PES: (L) foot, stalk; prob. fr. Sanskrit meaning pad or foot

There are two genera in the Pontederiaceae family: the *Heternanthera* and the *Eichornia*. The latter is represented by one species named after J. A. F. Eichorn, a Prussian minister of education. The plant is a native of tropical America but has become established as one of the world's worst floating aquatic weeds. Rarely a problem in cropped fields, it does its damage by competing for water before the water gets to the crop or by impeding flow to the field. It also seriously impedes navigation on many of the world's rivers and lakes. The specific name is derived from long, narrow, balloonlike, broad to nearly circular, thick, or perhaps even fat floating leaves. Leaf stalks are enlarged into basal, oval bulbs filled with air cells, which give the plant its buoyancy and its thick appearance.

Eleusine indica (L.)Gaertn. Goosegrass
(L′-u-scene in′-dih-ca)

ELEUSIS: (G) a town in Greece
INDICO: (L) to point out, indicate
INDICUM: (L) indigo, a blue-black color

The name *Eleusine* comes from the town of Attica in ancient Greece (actually the city-state of Athens) where Demeter, the god of agriculture, fertility, and marriage and Ceres, the goddess of harvests, were worshipped. It is a good name for a plant but one wonders why one with such an auspicious origin had to be assigned to this common, not very attractive weed of dryland and plantation crops throughout the tropics. This is not the type, or indicator, species for this genus, and like so many plants it is not its virtues or lack thereof but its appearance and general form that determine its classification. Linnaeus put *E. indica* in the genus *Panicum* and assigned the specific name due to the blue color at maturity. Joseph Gaertner, a German botanist, placed it in the genus *Eleusine* and assigned the specific name we now use.

Elodea canadensis L.C.Rich. Common Elodea
(L-o′-d-ah can-uh-den′-sis)

HELOS: (G) a marsh
HELODES: (G) marshy
CANAD: (L) fr. Canada
ENSIS: (L) suffix meaning belonging to; possessing

The long slender stems of this perennial weed of aquatic and marshy sites bear whorled leaves and solitary flowers. It is dioecious, with pistillate flowers normally appearing first. Fruits and seeds are rare because male plants are rare; reproduction is by fragmentation. At maturity both kinds of flowers grow to the water surface on the flower receptacle's extended, threadlike pedicel. Pollen floats free and randomly contacts pistillate flowers. The specific name indicates that it is a native of Canada.

Equisetum arvense L. Field Horsetail
(Eh-quih-c′-tum r′-vence)

EQUUS: (L) a horse
SETA: (L) bristle
ARVENS: (L) of the field
ARVUM: (L) arable field

***Equisetum palustre* L.** Marsh Horsetail

(Eh-quih-c′-tum pah-lus′-tray)

EQUISETUM: see above
PALUSTRE: (L) marsh or swamp

The complete etymology of the scientific name of this non-flowering perennial fern ally reveals Linnaeus' imagination. *E. arvense* reproduces by spores and spreads by creeping rhizomes. Two kinds of vegetative structures are observed: unbranched, fertile stems appear in spring, terminating in a spore-bearing cone that may be 2.5–10 cm long; sterile or vegetative stems appear later in the season as 0.3- to 1-m-tall jointed stems with hollow sections and whorls of four-angled branches. It is the sterile vegetative structure that looks like a horse's tail and thus makes sense of the generic name and yields its common name. One might call it the small bristle (tail) of the horse. It is a common field weed with a preference for wet places but has a capability of growing in many environments.

Equisetum palustre is a perennial, creeping fern ally without tubers but with rhizomes. Its appearance is similar to *E. arvense,* but this species demands a wet habitat.

***Eragrostis cilianensis* (All.)E.Mosher** Stinkgrass

(Ere-ah-gros′-tis sill-e-ah-nen′-sis)

ER: (G) fr. *earos* = spring
ERA: (G) the earth
EROS: (G) god of love
AGROSTIS: (G) the plant of the field; a grass
CILIOL: (L) hair or hairlike
ENSIS: (L) suffix meaning possessing; belonging to

The generic name has two possible derivations. The first reveals that this short, annual grass is a spring plant of the arable field. This is true and it is especially prevalent in areas of low rainfall. The second possibility is related to Eros, the Greek god of love. Members of the genus are collectively called love grasses. The common name of this grass is stinkgrass, apparently because of its offensive odor to cattle, which will not graze it. One expects the name love grass came from the generic name and not because this plant had anything to do with love. The leaf blades are 5–20 cm long and have many fine hairs on the upper surface, which makes them rough to touch. The underside and leaf sheath are smooth.

Erigeron strigosus Muhl. ex Willd. Rough Fleabane

(E-ridge′-er-on stri′-go-sus)

ER: (G) fr. *earos* = spring
ERI: (G) early
GERON: (G) old
STRIGA: (L) swath, furrow
OSUS: (L) a suffix meaning full of

The genus takes its name from the hoariness of some summer species. Stems and leaves are often covered with fine gray-white hairs. The hairs and the consequent gray-white appearance, i.e., the hoariness, give the impression of early old age. The specific name was assigned because of a pronounced groove created by the midrib of the lower leaves. Leaves of this species have a more pronounced groove than other members of the genus.

Erodium cicutarium (L.)L'Her. ex Ait. Redstem Filaree

(E-ro′-d-um sih-q-tare′-e-um)

HERODIOS: (G) a heron
CICUTA: (L) name of poison derived from hemlock
IUM: (L) dim. suffix

This annual or biennial herb is valued as spring forage in some places but can also be an aggressive weed. The genus was named because the mature capsule (seed pod) resembles the long bill of a heron. Why the heron was picked instead of another long-billed bird, we do not know. The specific name implies that the plant looks like a small *Cicuta* or poison hemlock. The resemblance is remote, at best, and is found in the pairs of leaflets that have toothed margins.

Eupatorium capillifolium (Lam.)Small Dogfennel

(U-pa-tor′-e-um cah-pill-ih-fo′-lee-um)

EUPATORIUM: (G) hemp agrimony
CAPILLUS: (L) hair, hairlike
FOLIUM: (L) leaf
IUM: (L) dim. suffix

Linnaeus thought that Eupator Mithridates, King of Pontus and known as Mithridates VI (died 63 B.C.), used one species of this genus for medicinal purposes. From him we have the word mithridatism, which is the production of immunity to poison by taking very small doses and, presumably, gradually building up immunity. Leaves of *E. capillifolium* are delicate, glabrous, and 2–10 cm long; the larger ones are pinnately divided into threadlike segments and give the plant its specific name.

Euphorbia esula **L.** Leafy Spurge

(U-for′-b-uh s′-u-la)

EU: (G) well, good
HORBE: (G) pasture
ESULA: (L) for eating

Euphorbia hirta **L.** Garden Spurge

(U-for′-b-uh her′-tah)

EUPHORBIA: see above
HIRTUS: (L) rough, hairy

One can derive the etymology of *Euphorbia* to mean good pasture but this is an unlikely origin. Several sources indicate that the generic name was given by Iuba, King of Mauritania, who named the herb after his physician Euphorbos because of his use of the herb. The specific name *esula* means the plant was for eating. Now, it is not eaten by humans but is by sheep, almost selectively, in a pasture, and they can be used as a biological control. It is a perennial weed with a milky latex, and it spreads by creeping roots and rhizomes. Stalks commonly occur in clumps, which can be easily distinguished because of the blue-green color of mature plants and their terminal yellow to green-yellow flowers.

The species *E. hirta* is distinguished by the dense mat formed on the soil surface by brown, crisp, hairy stems. Stems are prostrate and have the milky latex characteristic of this genus.

Fimbristylis littoralis **Gaud.** (syn. F. miliacea [L.]Vahl) Globe Fingerush

(Fim-brih-sty′-lis lit-or-al′-iss) (mill-e-a′-c-uh)

FIMBRIA: (L) fibers, threads
FIMBRIATUS: (L) bordered with a narrow band or edge of hairs
STYLIS: (L) of the style
LITTUS: (L) the seashore
ALIS: (L) pertaining to

LITTORALIS: (L) belonging to the seashore
MILIE: (L) thousands
ACEA: (L) a suffix meaning having, possessing

The type species of this genus has a style that is fringed with hairs exactly as the generic name says it should be. It is an erect annual or perennial sedge and is a common weed in aquatic environments, especially in lowland rice culture. *Miliacea* means thousands and refers to numerous, solitary, brown, or straw-colored, 2- to 2.5-mm diameter spikelets of the diffuse umbel inflorescence. *Littoralis* is consistent with the plant's aquatic habitat.

***Galega officinalis* L. Goatsrue**

(Ga′-lay-ga o-fis-in-al′-iss)

GALEGA: (L) literally, from Gaul meaning Gallic
GALEGA: (S) a plant name
OPIFICINA: (L) originally a storeroom or workshop, later pharmacy
OFFICINA: (L) an office

The name implies that the plant came from France or Northern Italy, which were once called Gaul and ruled by Rome. We know it was introduced into the United States in Utah in 1891 as a potential forage crop, but we do not know where it came from. The name could also be derived from *gala,* Latin for milk, but there is no plant-based reason to select this derivation. The specific name means that it was once used for its medicinal properties.

***Galinsoga parviflora* Cav. Small-flower Galinsoga**

(Gal-in-so′-ga par-vih-flor′-ah)

GALINSOGA: after Mariano Martinez de Galinsoga, a Spanish botanist, died 1797
PARVUS: (L) small, little
FLORA: (L) goddess of flowers
FLOS: (L) a flower

The Spanish botanist Antonio Jose Cavanilles (1745–1804) named this genus in honor of Galinsoga, another Spanish botanist. It grows 0.3–1 m tall and has numerous small, flowering heads. Ray flowers (four to five) are small, fertile, and white, while disk flowers are yellow.

Galium aparine L. Catchweed Bedstraw
(Gal′-e-um ap-par′-e-nay)

GALION: (G) bedstraw
GALA: (G) milk
APAIRO: (G) to take or seize

The generic name comes from the Greek word for bedstraw, the accepted common name in the United States, although it is called cleavers in Europe. The root word *gala* means milk and was assigned because at one time this species was used to curdle milk for making cheese. The slender, decumbent, trailing, four-angled stems have short bristly hooks on their edges. These hooks will readily attach to fur or a pant leg; thus, the specific name *aparine* is appropriate because they seize you as you pass by. It could also have been assigned because stems cling together or seize each other, and have been used to make a coarse strainer for milk that was curdled by the weed.

Grindelia squarrosa (Pursh)Dunal Curlycup Gumweed
(Grin-deal′-e-ah square-o′-sah)

GRINDELIA: after David Hieronymus Grindel, a Russian botanist, died 1836
SQUARROSE: (L) rough with scales

The erect stems of *G. squarrosa* are branched near the top, glabrous, and often tinged with red. The many-flowered heads are grouped in an open panicle with yellow ray flowers and closely imbricated green bracts; disk flowers are also yellow. The leaves have small spines and serrated edges and feel rough or squarrose when touched. The most striking feature of this weed is represented by its common name, gumweed. The flowers, especially near involucral bracts, are sticky. This may be a method of seed dispersal or a way to attract insects.

Gutierrezia sarothrae (Pursh)Britt. & Rusby
Broom Snakeweed
(Goo-t-air-e′-z-uh sah-ro′-thray)

GUTIERREZIA: after Pedro Gutierrez, a correspondent of the botanical garden of Madrid
SAROTHRON: (G) broom
SAREIN: (G) to sweep

Pedro Gutierrez, a member of a Spanish noble family, was honored by having this genus named after him. The specific name is descriptive of the numerous herbaceous leafy branches, which shed the lower leaves at maturity. The ascending, hairless, and ridged branches were probably used as brooms although they bear only a superficial resemblance to brooms today.

Halogeton glomeratus **(Stephen ex Bieb.)C.A.Mey.**

Halogeton

(Hal-o-g′-ton glom-er-a′-tus)

HALOS: (G) the sea
HALO: (G) to breathe
GEITON: (L) neighbor
GLOMUS: (L) a ball, a sphere, round body
ATUS: (L) suffix meaning provided with

This annual plant, imported to North America from Siberia, can be as much as 1 m in diameter and 0.3 m tall. It is often much smaller because it commonly occurs on poor, dry, saline rangeland soils. It is not only a neighbor of the sea and saline areas, it lives in saline soil. *Halogeton* is an aggressive invader and poisonous to sheep. The specific name might be related to the cross-sectional shape of sausagelike, inflated leaves. It is much more likely that it is related to the fact that flowers are borne in glomerules or compact clusters. Inconspicuous yellow-green flowers are borne with two to several in a cluster in leaf axils. There are two flower types on one plant. Showy flowers have large membranous sepals, which spread fanlike above a slight constriction at the seed apex. The entire structure is spheroid. The second flower type has much-reduced sepals but retains a spheroid outline.

Hedyotis corymbosa **(L.)Lam.** Hedyotis

(Hed-e-o′- tus cor-im-bo′-sa)

HEDYLOS: (G) sweet, pleasant to taste or smell
OTIS: (L) prob. fr. *itis* = like
CORYMBUS: (L) a cluster of flowers; now a flat-topped or merely convex open flower cluster
KORYS: (G) helmet

This erect or decumbent annual is a tropical weed. The generic name is not particularly appropriate for it does not have an especially pleasant odor and its taste is debatable though not offensive. There are no records of its being eaten because of its sweet taste. The flowers do fit the botanical definition of a corymb, which is a flat or convex flower cluster with outer flowers opening first. It usually flowers with two to five umbels arising from leaf axils along the stem; some solitary flowers are intermixed. The flowers are white, small, and each may be perceived to be covered by a helmet of short, pointed calyx lobes. Two-celled fruit capsules do not protrude beyond these lobes.

Helianthus petiolaris **Nutt.** Prairie Sunflower

(He-lee-an′-thus pet′-e-o-lair-iss)

HELIOS: (G) sun
ANTHOS: (G) a flower
PETIOLUS: (L) a little foot, stem or stalk
ARIS: (L) suffix meaning pertaining to

Sunflowers look like a sun with their dark center disk flowers and radiating yellow ray flowers. They also follow the sun in its daily path across the sky. The plant has large, coarse, mostly entire leaves that alternate on the stem. The entire plant and its flowers are supported on stems as the name says they should be.

Heliotropium indicum **L.** Indian Heliotrope

(He-lee-o-tro′-p-um in′-dih-cum)

HELIOS: (L) sun
TROPIUM: (L) following
TROPOS: (G) a turn
TREPO: (G) to turn
INDICUM: (L) the color indigo, a dark blue dye

Some species of *Heliotropium* are perennials or biennials; this one is an annual. It is not as conspicuous a sun follower as is *Helianthus petiolaris.* The inflorescence is usually a series of terminal narrow spikes, each 15–20 cm long, which curve backward at the tip and have numerous small white, lavender, or purple flowers on the upper side of the spikes in two closely packed rows. Those originally identified were very likely purple and gave the plant its specific name.

Hibiscus trionum **L.** Venice Mallow

(Hi-bis′-cus tri-o-num)

HIBISKOS: (G) mallow of the marsh; obscure origin
TRIONUM: (G) in three parts

Hibiscus has an obscure origin. It may have been a plant of the marsh, and it is a member of the mallow family. The alternate, simple, pinnately veined, petioled upper leaves are divided into three distinct lanceolate, pinnate lobes.

Hordeum jubatum **L.** Foxtail Barley

(Hor′-d-um jew-bay′-tum)

HORDEUM: (L) barley
JUBA: (L) a mane
JUBATUS: (L) having a mane or fringe of hairs; crested

Hordeum is the Latin name for barley and Linnaeus chose the Latin *jubate* because he saw a resemblance between the long-awned, nodding, 5- to 12-cm-long, jointed spikes and a horse's mane.

Hydrilla verticillata **(L.f.)Royle** Hydrilla

(Hi′-dril-la ver-tih-sih′-la-ta)

HYDOR: (G) fr. *hydratos* = water
ILLA: (G) used to form adverbs; means by that way or there
ILLO: (G) to roll, turn
ILLODES: (G) distorted
VERTICIL: (L) whorl, whorled
ATA: (L) suffix indicating possesion or likeness

This small water weed is undesirable in water. As the generic name implies it is an aquatic species, but the reason for the choice of the suffix is not clear. It could have been chosen because it lives in water, but it could also be because it seems to roll and turn in water and appears distorted. Its dense, long, branching stems form large floating mats. The specific name comes from the fact that the leaves occur in whorls at nodes.

Hypericum perforatum L. Common St. Johnswort

(Hi-per′-ih-cum per-for-a′-tum)

HYPER: (G) over, above, or on a heath
HYPERION: (G) the sun god
EIKON: (G) a picture, an image
YPER: (G) upper
ERIKE: (G) a plant of the heath, or the heath
PERFORO: (L) to pierce through

The derivation of the generic name means a plant over, under, on, or of the heath. A heath is often a wasteland or, at best, open uncultivated land, usually with poor drainage and coarse soil and a high content of peat or peaty humic material that will burn. The generic name comes from the fact that this plant's flowers were often placed above religious images to ward off evil spirits at the midsummer festival of *walpurgisnacht,* which became the feast of the nativity of St. John the Baptist, celebrated on June 24th. We do not know why this plant was chosen. It may have been one of the few in bloom at that time in the area where it was used. The origin of the common name is obvious from this story, and the use of the suffix "-wort" implies that it was used as a medicinal plant. The appropriateness of the specific name will be seen when one holds the leaf up to light and sees translucent spots on each leaf; however, they are not perforations.

Imperata cylindrica (L.)Beauv. Cogongrass

(Im-per-ah′-ta sih-lin′-drih-cuh)

IMPERATA: after Ferrante Imperato, an Italian apothecary, died 1625
KYLINDROS: (G) a cylinder
CYLINDRICA: (L) a cylinder, cylindrical

Linnaeus named this genus of mostly tropical grasses after the Italian naturalist and apothecary Imperato. This species is a perennial grass, which grows 15–120 cm high from seed and from an extensive system of rhizomes. The specific name is undoubtedly related to the dense, white, fluffy, cylindrical, spikelike panicles. They are 3–20 cm long (rarely longer) with obvious creamy white, silky hairs.

Ipomoea hederacea **(L.)Jacq.** Ivyleaf Morningglory

(Ip-o-me′-ah hed-er-a′-c-ah)

IPS: (G) worm
IPOS: (L) a worm that eats vines or wood
OMOEA: (L) suffix meaning like
HOMOIOS: (G) like
HEDERA: (L) ivy
GHED: (L) to clasp or hold

Ipomoea pes-tigridis **L.**

(Ip-o-me′-ah pes-tig′-rid-iss)

IPOMOEA: see above
PES: (L) foot, stalk; prob. fr. Sanskrit meaning pad or foot
TIGRIDIS: (L) fr. *tigris* = tiger

Ipomoea purpurea **(L.)Roth** Tall Morningglory

(Ip-o-me′-ah purr-purr′-e-uh)

IPOMOEA: see above
PURPUREA: (L) purple

Ipomoea triloba **L.** Threelobe Morningglory

(Ip-o-me′-ah tri′-lo-ba)

IPOMOEA: see above
TRI: (L) three
LOBA: (L) lobed

Anyone familiar with the twisting, climbing, aggressive nature of morningglory will see that the generic name is appropriate. Wiggly and undulating, like a worm, describes how these plants behave in the field, except they are longer and more aggressive than worms. Leaves are alternate, simple, palmately veined, and three-lobed; the flowers are perfect, regular, and showy. The funnel-form corolla is 1–2 in. long and white, purple, or pale blue. An annual, *I. hederacea* twines or trails along the ground, and when the stem reaches something it can climb, it does. It appears like ivy as it clasps and twines up another plant or a fencepost.

The distinctive characteristic of the annual vine *I. pes-tigridis* is its leaves. They are 6–10 cm in diameter, palmate, and distinctly lobed with from five to nine lobes. Most of us have not had the chance to investigate a tiger's foot or pad, but our imagination tells us that it probably looks like this leaf.

I. purpurea can have vines up to 3 m long. The specific name tells us that it has predominantly purple flowers that look like a typical morning-glory flower. However, flower color is not a positive identification because while they are predominantly purple, they can also be white to light blue.

The name *I. purpurea* tells us to look for the color purple. The specific name *triloba* tells us that a key identifying trait is something that has three lobes. The *I. triloba* seedling always has three-lobed leaves, but mature plants generally have broadly ovate, entire, coarsely dentate or three- to five-lobed leaves 2–10 cm long. Again we have a good name that aids identification but is not an absolute guide.

Ischaemum rugosum **Salisb.** Saramollagrass

(Iss'-sky-mum rue-go'-sum)

ISCH: (G) to check
ISCHEIN: (G) to check, restrain, or suppress
ECHEIN: (G) to have or hold
ISCHAIMOS: (G) to contract, styptic
ISCHI: (G) fr. *ischion* = hip joint
ISCHNOS: (G) slender, withered
RUGOSUS: (L) wrinkled
RUGA: (L) wrinkle

The origin of *Ischaemum* is not clear. One commonly finds reference to the fact that *Andropogon ischaemum* L. was a styptic plant, which means that at one time it was used to check blood flow; this is consistent with some possible origins of the generic name. The weed *I. rugosum* has no known styptic properties, but they may have been lost with changing medical practices. One also finds that the Greek *ischi* refers to the mammalian hip joint. There is a chance the name was chosen because the inflorescence is apparently single when young but separates at maturity into two spikelike racemes, each 5–10 cm long. Finally one has to admit ignorance about why this name was chosen and how it relates to the plant but it may have something to do with the long, slender stems. The yellow-green lower glume has prominent transverse ridges and is thus rugose, or wrinkled, the origin of the specific name.

Kochia scoparia **(L.)Schrad.** Kochia

(Ko'-shah sco-pair'-e-uh)

KOCHIA: after W. D. J. Koch, a German botanist, died 1849
SCOPULA: (L) thin branches, twigs
SCOPARUM: (L) a broom

This large, bushy, annual weed could function as an effective broom when green or after senescence. It would not be as efficient as our modern brooms but it would work and probably was used in the past.

Lactuca serriola **L.** Prickly Lettuce

(Lak′-too-kah serr′-e-o-la)

LACTUCA: (L) lettuce
LAC: (L) milk
SERIOLA: (L) serrated, sawlike

Lactuca is Latin for lettuce, and although there is little outward resemblance between this plant and garden or leaf lettuce in its seedling or mature stage, they are related. The specific name fits well. The alternate leaves are simple and clasping. The underside of the white midrib has a row of short spines that resemble saw teeth.

Lamium amplexicaule **L.** Henbit

(Lam′-e-um am-plex-ih-caw′-lay)

LAMION: (G) dim. of *lamia* = monster
LAMINA: (L) a thin plate
LAIMOS: (G) throat or gullet
AMPLECTO: (L) to entwine or encircle
KAULOS: (G) stem or stalk

I suppose if one has this plant in a lawn it could be regarded as a monster, but it is unlikely that this is the correct etymology. The name may come from the thin, palmately lobed, petioled lower leaves. The most likely etymology is the tubular, pink to purple throatlike corolla of flowers borne in whorls in leaf axils. The basal portion of upper sessile leaves clasps the stem and accounts for the specific name.

Lantana camara **L.** Largeleaf Lantana

(Lan-tan′-uh cam-r′-uh)

LANTANA: (L) prob. fr. Italian; or related to viburnum
KAMARA: (G) a vaulted chamber
CAMARA: (L) to bend or curve; also, a room
CHAMBRE: (F) a room with a vaulted roof
CAMARA: (S) vernacular for a species of lantana

The origin of *Lantana* is probably from the Italian name viburnum, which this plant's foliage resembles. It is a spreading, aromatic, perennial, thicket-forming shrub with variably colored, quite pretty, small flowers. The specific name *camara* is of equally obscure origin. In Latin and modern Italian *camara* is translated as room, but the relationship to the plant is not clear. The best guess is that it is related to the ovoid, 4- to 6-mm diameter fruit borne on a thickened, fleshy receptacle. The fruit is green when immature but turns black or purple at maturity and contains two seeds.

Leersia hexandra **Sw.** Southern Cutgrass

(Leer′-z-uh hex-an′-drah)

LEERSIA: after Johann D. Leers, a German botanist, died 1774
HEX: (L) six
ANDRA: (L) refers to stamens

This is a tall, slender, leafy, perennial grass. On careful investigation one will find six stamens in each flower of the terminal panicle.

Lemna minor **L.** Common Duckweed

(Lem′-nuh mi′-nor)

LEMNA: (G) a water plant; origin uncertain
MINOR: (L) smaller, lessor, inferior

There are about 15 species of *Lemna* worldwide. They are all minute, floating aquatic plants, and this is one of the smaller ones, hence the specific name.

Lepidium campestre **(L.)R.Br.** Field Pepperweed

(Leh-pid′-e-um cam-pes′-tray)

LEPIS: (G) scale, flake
IUM: (G) dim. suffix
LEPISMA: (G) scale or rind
CAMPESTER: (L) of the fields, growing in fields or open country

Once called poor man's pepper, *L. campestre* is an annual or biennial weed in many different plant environments. The generic name is derived from a small scale, which covers the seed. Each seed is contained in a broadly ovate pod, which is winged all around with the style extending through a narrow notch at the top. The seed pod may have looked scaly to the namer.

Leptochloa chinensis **(L.)Nees** Chinese Sprangletop

(Lep'-toe-clo-ah chi-nen'-sis)

LEPTOT: (G) thin, fine, slender
LEPTOTES: (G) thinness
CHLOE: (G) young grass
CHIN: (L) fr. China
ENSIS: (L) suffix meaning place of origin

This annual grass grows in tufts or bunches. The stems are quite slender and ascend from a branching base. It is their slender appearance and the long attenuated inflorescence that led Linnaeus to choose *Leptochloa*. Its place of origin is tropical Asia, perhaps China, as the specific name indicates.

Linaria vulgaris **Mill.** Yellow Toadflax

(Lin-air'-e-uh vul-gare'-iss)

LINUM: (L) flax; thread
LINEUS: (L) flaxen, of flax
LINARIUS: (L) a linen weaver
VULGATUS: (L) ordinary, common
VULGARE: (L) to make known
VULGUS: (L) mob, common people

Many weedy species of *Linaria* resemble flax because of leaf similarities. *L. vulgaris* is the common or ordinary species, which is a perennial and it has very attractive flowers like snapdragons. It may have been called *vulgaris* because it was found everywhere or because it was commonly found in association with humans and thus was made known.

Lindernia anagallidea **(Michx.)Pennell** False Pimpernel

(Lin-dern'-e-uh an-ah-gah-lid'-e-uh)

LINDERNIA: after Johann Linder, a Swedish botanist, died 1723
ANAGALEO: (G) to decorate
ANAGELOS: (G) to laugh
ANA: (G) again
AGALLEIA: (G) to delight in
ANAGALLIS: (G) a plant, prob. pimpernel or a chickweed

Lindernia pusilla (Willd.)Bolding

(Lin-dern′-e-uh poo′-sih-la)

LINDERNIA: see above
PUSUS: (L) small child
PUSILLUS: (L) very small, weak

It is doubtful that the specific name *anagallidea* has anything to do with laughing or humor. The corolla is blue-violet and although small, it is pretty. The linkage of decoration or delight with its beauty is appropriate.

The specific name *pusilla* reveals that this is one of the smaller members of the genus.

Lolium multiflorum Lam. Italian Ryegrass

(Lo′-lee-um mul-tih-flor′-um)

LOLIUM: (L) darnel; origin uncertain
LOLLI: (L) to let hang
MULTI: (L) many
FLORA: (L) flower

Lolium temulentum L. Poison Ryegrass

(Lo′-lee-um tem-u-len′-tum)

LOLIUM: see above
TEMULENTUS: (L) intoxicated, nodding; top-heavy

Lolli means to let hang but it doesn't fit this plant. The more likely etymology is the uncertain origin of the Latin *darnel,* which is a frequent common name for ryegrass. The spikes are many-flowered and each spikelet has 10–12 flowers, more than other species of *Lolium.*

Lolium temulentum was introduced to the United States from Europe. It is a stout, erect annual with smooth stems, each 0.3–l m high and bearing five to seven leaves. Some authors suggest that it is the tares spoken of in the Bible (Matthew 13:25). The specific name refers to the poisonous nature of the seeds caused by a fungus (*Stromatinia temulenta* Prill.&Del.), which possesses a poisonous alkaloid, temulin.

Ludwigia adscendens (L.)Hara

(Lud-wih′-g-uh ad-sen′-dens)

LUDWIGIA: after Christian G. Ludwig, a German botanist, died 1773
ADSCENDENS: (L) ascending
ADSCENDO: (L) to ascend, mount up, to climb

Ludwigia octovalvis (Jacq.)Raven
Long-fruited Willow Primrose
(Lud-wih′-g-uh ock-toe-val′-vis)

LUDWIGIA: see above
OCTO: (L) eight
VALVA: (L) the leaves or folds of a door

Ludwigia adscendens seems to be misnamed because it does not grow anywhere near as tall as does *L. octovalvis,* which grows up to 1.5 m high. In fact, *L. adscendens* is a decumbent plant that roots in the mud of rice paddies and floats on water. The stems can be long, but they do not ascend.

Available botanical descriptions show that *L. octovalvis* is a stout, sometimes woody or shrubby annual weed, which has a cylindrical to somewhat club-shaped fruit (a capsule). The capsule is four celled with eight ribs and several rows of seeds in each cell. It is the eight ribs that are the origin of the specific name.

Lychnis alba Mill. (syn. *Silene alba* [Mill.]E.H.L.Krause)
White Campion
(Lich′-nis al′-ba) (Sigh′-leen)

LYCHNOS: (G) a lamp, a light
LEUKOS: (G) bright, white
ALBA: (L) white

The generic name *Lychnis* is a Greek term used to refer to the brightly colored flowers. Each plant is dioecious, and the flowers are solitary on long peduncles or they may occur in cymose clusters. The flowers open in the evening with five white or pink petals extending above the fused calyx. The name could also be from use of the thick, cottony substance on leaves of some species as wicks for oil lamps. The origin is not clear. See also *Silene alba*.

Lysimachia punctata L. Dotted Loosestrife
(Li-sih-mock′-e-uh punk′-tay′-ta)

LYSIMACHUS:(G) after Lysimachus, a 5th or 4th century B.C. Greek physician
PUNCTUS: (L) a stinging; a puncture
PUNCTATA: (L) ending in a point
PUNCTUM: (L) point
ATUS: (L) suffix indicating possession or likeness

The genus was named in honor of Lysimachus by Linnaeus. The specific name probably refers to the glandular margins of corolla lobes. This leafy-stemmed perennial has opposite or whorled leaves and yellow flowers. Members of this genus also commonly have glandular dots or points on their leaves.

Lythrum salicaria **L.** Purple Lythrum

(Li′-thrum sal-ih-care′-e-uh)

LYTHRON: (G) blood, gore; often used to denote red color
LYTRON: (G) a name used by Dioscorides for this species; like a willow
SALIX: (L) the willow
ARIA: (L) suffix meaning like or connected with

The generic name probably came from the pink to magenta (rarely white) flowers. It is also possible that the plant or some portion of it has styptic properties and was used to stop bleeding. This annual plant frequently grows in marshes and along shores of lakes and rivers. The specific name comes from an observed resemblance of the opposite, or whorled (usually in threes), sessile, lanceolate to narrowly oblong leaves, which are cordulate or heart shaped at their base, to the linear shape of willow (genus *Salix*) leaves.

Malva neglecta **Wallr.** Common Mallow

(Mal′-va neh-glec′-tuh)

MALVA: (L) the mallow
MALAKOS: (G) soft or soothing
MALASSO: (G) soften
NEGLECTUS: (L) neglected
NEC: (L) not
LEGO: (L) to choose or gather

The Latin *malva* came from the Greek in which it meant soft or soothing. This was probably an allusion to an emollient derived from seeds or leaves. It could also have been given because of the soft leaves or its reputed relaxing powers when taken as a tea. *Neglecta* is a rather good name for a weed because when it is neglected, it will soon gain your attention, for it takes over your lawn or garden. It is more likely that it was a plant not chosen or gathered for food, an origin consistent with its etymology.

Marsilea minuta L. Water Clover
(Mar-sill′-e-ah mih′-new-ta)

MARSILEA: after Count Luigi Fernando Conte Marsigli, an Italian naturalist, died 1730
MINUTA: (L) small, finely divided
MINUO: (L) to make small

This is one of the smallest members of a genus of small perennial ferns. The leaves resemble leaves of clover and it is commonly called water clover.

Matricaria inodora L. (syn. *M. perforata* Merat) Scentless Chamomile
(Mat-trih-care′-e-uh in-o-dor′-ah) (per-for-ah′-ta)

MATRICIS: (L) womb or uterus
MATRIC: (L) often used to mean place of origin
MATRICARIA: (L) feverfew, a plant
ARIA: (L) suffix meaning pertaining to or related to
ODORUS: (L) fragrant
IN: (L) negative
INODORUS: (L) having no smell, scentless
PERFORATUS: (L) pierced, with holes

Matricaria maritima L. (syn. *Tripleurospermum maritimum*[L.]W.D.J.Koch) False Chamomile
(Mat-trih-care′-e-uh-mare-ih-t′-ma) (Triple-u-ro-sper′-mum mare-ih-t′-mum)

MATRICARIA: see above
MARITIMUS: (L) growing by the sea

Matricaria matricarioides (Less.)C.L.Porter Pineappleweed
(Mat-trih-care′-e-uh mat-trih-care-e-oy′-deez)

MATRICARIA: see above
OIDES: (G) resembling or like

Members of this genus were used in folk medicine to treat menstrual disorders and this is the most probable origin of the generic name. The Latin name *matric* comes from *matricis,* the uterus or womb, and is often used to mean a place where anything is generated. Some members of this genus have an odor but *M. inodora* is unique because it lacks odor. There is no apparent basis for the new specific name *perforata* because no plant part is pierced with holes or has pores.

Matricaria maritima can surely grow by the sea, but it is not bound to do so and is common on many agricultural sites. See also *Tripleurospermum maritimum*.

The species *M. matricarioides* is a much-branched annual plant without hairs that grows 6–18 in. tall and has alternate leaves. The leaves have a distinct pineapple odor when crushed. There are numerous yellow-green disc flowers grouped into a cone-shaped head about 1/4 in. in diameter. The plant was probably considered the type species for the genus, which explains its specific name.

Mikania cordata **(Burm.f.)B.L.Robins.** African Mile-a-minute

(Mi-cane'-e-uh cor'-da-ta)

MIKANIA: after Joseph Mikan, a Czech botanist, died 1814
CORDIS: (L) the heart
ATUS: (L) suffix indicating possession or likeness

The specific name refers to the general heart shape of the leaves of this perennial twining vine.

Mimosa invisa **Mart.** Giant Sensitive Plant

(Mih-mo'-sa in'-v-sa)

MIMULUS: (L) a mimic, actor, mime
INVISUS: (L) hated, detested, also overlooked or unseen

Mimosa pudica **L.** Sensitive Plant

(Mih-mo'-sa poo'-dih-cah)

MIMOSA: see above
PUDICUS: (L) shamefaced, bashful, modest

Mimosa comes from the Latin for mimic or mime. The giant sensitive plant is not as good a mimic as is *M. pudica*. When any exposed part of *M. pudica* is touched, the 12–25 pairs of leaflets on one or several leaves fold up very quickly, whereas *M. invisa* is only moderately sensitive to touch. Thus, these plants, by covering themselves when touched, mimic bashful or modest behavior. It took a good imagination and knowledge of Latin to find these words to clearly describe the plant's behavior rather than its looks. It has characteristic, recurved spines and was probably not an overlooked plant; however, it was likely detested by those who had it.

Mollugo pentaphylla L.

(Moll'-oo-go pen-ta-fill'-ah)

MOLLIS: (L) soft, mild
PENTA: (G) five
PHYLLA: (G) leaves

Mollugo verticillata L. Carpetweed

(Moll'-oo-go ver-tih-sih-la'-ta)

MOLLUGO: see above
VERTICIL: (L) whorl
VERTO: (L) to turn
ATA: (L) suffix indicating possession or likeness

Mollugo comes from Latin and means soft or mild and refers to no aspect of these plants. Softness is most often imparted by pubescence, and these plants are glabrous throughout. One member of the genus undoubtedly was perceived to be soft by Linnaeus or had now unknown emollient properties. *Pentaphylla* refers to the five leaves in each whorl of leaves on the stem. In the species *verticillata* there are three to six leaves in each whorl.

Monochoria vaginalis (Burm.f.)Kunth Monochoria

(Mon-o-core'-e-uh vaj-in-al'-iss)

MONO: (L) one
CHORIA: (L) chord or string
VAGINA: (L) a scabbard, sheath; the covering sheath holder
ALIS: (L) pertaining or related to

This fleshy, semiaquatic monocotyledonous weed looks like a dicot but is not. The generic name comes from the fact that each flower has only one stamen. Each inflorescence is basally opposite the sheath of a floral leaf. This position gives rise to the specific name. From 3 to as many as 25 violet or lilac-blue flowers arise as a 3- to 6-cm spike from a thickened bundle about two-thirds of the way up the leaf stalk.

Murdannia nudiflora (L.)Brenan

(Mur-dan'-e-uh new-dih-flor'-uh)

MURDANNIA: unknown origin
NUDI: (L) bare, naked
NUDUS: (L) naked, stripped of cover
FLORA: (L) flower

This weed is related to *Commelina benghalensis* and *C. diffusa,* which are annuals or perennials of moist places in the tropics and subtropics. The flowers are not especially naked. They are about 6 mm in diameter and numerous in short-lived racemose clusters at the ends of few branches. Flowers occur on 3- to 7-cm-long terminal peduncles, sometimes with a single bract, and may therefore be regarded as stripped of cover.

Opuntia polyacantha **Haw.** Plains Pricklypear

(O-pun′-t-uh polly-ah-can′-thah)

OPUNTIA: (L) a plant from Opountos, Greece
OPUS: (G) a plant, but not this one
POLY: (G) many, numerous
KANTHANOS: (G) a deep cup with a stem and looped handle
CANTHARUS: (L) large, wide-bellied drinking vessel

This weed of rangeland and dry arid lands may have occurred in Greece and may still grow near the ancient city of Opountos in the Locris area. *Opuntia* was used by Pliny and Theophrastus, but it is not possible to discern any other etymology that relates to looks or behavior. The specific name could stem from the many cup-shaped yellow to red flowers that occur at the ends of pads. It could also be related to the fact that its pads can serve as a reservoir for water, yet the reservoir is not external but internal and pads cannot be regarded as cups. A view (or feel) of their spines would quickly convince one of the lack of utility as a cup; however, when the spines are removed pads can be eaten by cattle.

Orobanche crenata **Forsk.** Crenate Broomrape

(Or-o-bank′-e cren′-ah-ta)

OROBOS: (L) bitter vetch
ANCHEIN: (L) to strangle
CRENATE: (L) having rounded teeth
CRENULA: (L) a notch
CRENATUS: (L) notched

The genus *Orobanche* is composed of annual, true parasitic plants that live on roots of several important crops. They represent a bitter vetch that strangles a plant by removing its nutrients. The crenate feature of this species is a series of rounded teeth on margins of reduced leaves.

Oxalis corniculata L. Creeping Woodsorrel

(Ox′-ah′-lis core-nik-u-la′-ta)

OXALIS: (L) garden sorrel
(G) sorrel
OXUS: (G) acid, sour
OXYS: (G) sharp, keen, pointed
CORNU: (L) a horn
CORNIS: (L) horned
CORNICULUM: (L) small hornlike appendage
ATA: (L) suffix indicating possession or likeness

Oxalis comes from Latin and means garden sorrel. Its ultimate etymology is Greek and means acid or sour; the leaves are acid and taste sour. The specific name was given because the oblong fruit capsule often stands up and appears to be horned. The 1- to 1.2-mm-long, egg-shaped seed has a point or small horn on one end.

Oxytropis sericea Nutt. ex T. & G. Silky Crazyweed

(Ox-e-trope′-iss sir-e′-c-uh)

OXUS: (G) acid, sour
OXYS: (G) sharp, keen, pointed
TROPIS: (G) keel
TREPEIN: (G) to turn
SERIKON: (G) silk
SERIS: (L) silk
SERICUM: (L) silky, silk

This poisonous perennial herb arises from a woody taproot and has a much-branched crown with all basal leaves. The racemose flowers are congested on a peduncle. Flowers are typical of the legume family with a corolla composed of an upright, more or less erect banner petal, two lateral wing petals, plus a keel of two united petals. These united petals form a pointed keel that terminates in a forward bending beak. The entire plant is covered with fine, silky hairs establishing the correctness of the specific name.

Panicum capillare L. Witchgrass
(Pan′-ih-cum cap′-ill-air)

PANUS: (L) ear of millet
PANIS: (L) bread
PANICUM: (L) millet
(L) anything baked, bread, cakes, etc.
CAPILLUS: (L) hair

Panicum dichotomiflorum Michx. Fall Panicum
(Pan′-ih-cum di-cot-o-mih-flor′-um)

PANICUM: see above
DI: (L) two
DICHA: (G) in two
TEMNEIN: (G) to cut
DICHOTOMEIN: (G) to cut in half
FLORUM: (L) flower
FLOS: (L) flower

Panicum maximum Jacq. Guineagrass
(Pan′-ih-cum max′-ih-mum)

PANICUM: see above
MAXIMUM: (L) greatest, most

Panicum miliaceum L. Proso Millet
(Pan′-ih-cum mill-e-a′-c-um)

PANICUM: see above
MILIACEUM: (L) consisting of millet
MILIARIUS: (L) consisting of millet
MILIUM: (L) millet

Panicum repens L. Torpedograss
(Pan′-ih-cum ree′-pens)

PANICUM: see above
REPENS: (L) creeping
REPO: (L) to creep

The generic name *Panicum* probably was given either because some of these species have been and still are used to produce grain or because seed from some species is used to make flat bread. *P. capillare* has a large open panicle that looks hairy, but the specific name is a reference to hairiness of the leaf sheath.

When Andre Michaux (1746–1802) named *P. dichotomiflorum* he saw, or thought he saw, that the flowers occurred in pairs on two equal branches. However, the flowers are not always in pairs nor are they always on equal branches, but the name remains.

To be maximum is to be large, and *P. maximum,* a perennial grass, grows up to 4 m tall so the specific name is appropriate.

P. milaceum is an annual and a close relative of cultivated millet, which bears the same scientific name. It escaped to become a weed in many central U.S. states. The specific name means that it resembles millet, which it does. This may be an example of a plant where continued taxonomic investigation will result in further refinement of identification, and it may become a separate species.

P. repens is well named because it is an erect perennial that spreads via long, stout rhizomes.

***Paspalum conjugatum* Bergius** Sour Paspalum

(Pass-pal′-um con-jew-gay′-tum)

PASPALOS: (G) millet
CONJUGATUS: (L) joined, connected

***Paspalum dilatatum* Poir.** Dallisgrass

(Pass-pal′-um dill-uh-tay′-tum)

PASPALUM: see above
DILATO: (L) to dilate
DILATUM: (L) broadened, expanded
DILATATUS: (L) widened

***Paspalum distichum* Am. auctt.** (syn. *P. paspaloides* [Michx.]Scribn.) Knotgrass

(Pass-pal′-um dis′-tih-cum) (pass-pal-oy′-deez)

PASPALUM: see above
DISTOICHOS: (L) consisting of two rows
PASPALOS: see above
OIDES: (L) resembling

Members of the genus *Paspalum* are mostly perennials and all are grasses. The inflorescence is a panicle, and earlier they were probably classed with members of the genus *Panicum* until more careful taxonomic work separated them. Thus, we have two genera where the name can be traced back to a Greek reference to millet. *Paspalum conjugatum* is a creeping, prostrate perennial that grows up to 60 cm tall. The inflorescence is the source of the specific name. It has two (rarely three) terminal, spike-like, very slender branches with two rows of yellow, approximately circular, overlapping and thus conjugated, or joined, spikelets along the underside of a flattened axis.

P. dilatatum also obtains its name from the inflorescence. Paired spikelets are basally rounded and quite flattened compared to other *Paspalum* species. One has to compare several species to see how appropriate the name is.

P. distichum has two (rarely three and sometimes one) terminal, erect spikes similar to *P. conjugatum*. Spikelets are arranged in two definite rows along one side of the spike. *P. paspaloides* is the new and preferred name.

***Pennisetum clandestinum* Hochst. ex Chiov.** Kikuyugrass
(Peh-nih′-c-tum clan-des′-tih-num)

PENNI(A): (L) feather, wing
SETUM: (L) bristle
SETA: (L) bristle
CLANDESTINUS: (L) secret, hidden, concealed

***Pennisetum pedicellatum* Trin.** Kyasumagrass
(Peh-nih′-c-tum peh′-dih-cell-a-tum)

PENNISETUM: see above
PEDICELL: (L) pedicelled, having pedicells
ATUM: (L) suffix meaning provided with

***Pennisetum polystachon* (L.)Schultes** Missiongrass
(Peh-nih′-c-tum poly-stay′-chon)

PENNISETUM: see above
POLY: (G) many
STACHON: (G) relating to a spike
STACHYS: (G) an ear of grain

***Pennisetum purpureum* Schumach.** Napiergrass
(Peh-nih′-c-tum purr-purr′-e-um)

PENNISETUM: see above
PURPUR: (L) fr. *purpuratus* = clad in purple
EUM: (L) suffix denoting place

Pennisetum refers to the inflorescence, which is a terminal, dense, cylindrical spike. Each spikelet of this perennial weed is surrounded by bristles giving the appearance of a bristly wing or feather. *P. clandestinum* is well named because it produces few inflorescences that are well hidden in the uppermost leaf sheath. Stamens are the only visible flower parts and they appear as a mass of fine, white threads attached to leaves.

The specific name for each of the other plants in the genus listed here fits well. In the perennial *P. pedicellatum* each spikelet sits on a small pedicel or stalk. The many spikes of the annual *P. polystachon* form a single tussock or clump, which characterizes this species. *P. purpureum* obtains its specific name from numerous, dark yellow, brown, or purple bristles that subtend each spikelet. They are not all purple although that is apparently what Heinrich C. F. Schumacher (1757–1830) of Denmark saw.

***Phalaris arundinacea* L.** Reed Canarygrass

(Fa-lair'-iss ah-run-din-a'-c-uh)

PHALARIS: (L) canarygrass
PHALAROS: (L) having a white spot
PHALIOS: (G) having a white spot
PHALOS: (G) shining; white
ARUNDIN: (L) a reed
ACEA: (L) resembling

Phalaris means canarygrass in Latin and this perennial was once used to produce seed for canaries and other birds. It escaped to become an important weed of aquatic sites. The use of the term that means a white spot may be related to one species, *P. picta,* in which leaf blades are striped with white. The seed is shiny and straw colored. It is also possible the name relates to the seed when free of its coat, which hides shiny translucence. The tall 60- to 150-cm culms resemble reeds.

***Phragmites australis* (Cav.)Trin. ex. Steud.** Common Reed

(Frag-mi-teez aus-tray'-lis)

PHRAGMOS: (G) screen, partition, fence, or hedge
ITES: (G) suffix meaning pertaining or related to
AUSTRALIS: (L) south or southern

This perennial grass occurs worldwide in fresh water and may be the most widely distributed of all angiosperms. It grows stout leafy culms from 1.8 to 3.7 m tall, and they can form a screen or partition. Because it is so strong and tall the stems have been used to weave mats, screens, and lattices for construction of adobe and fiber structures. The specific name implies southern, not Australian, origin.

Phyllanthus debilis Klein ex Willd.

(Fi-lan′-thus d′-bill-iss)

PHYLLON: (G) leaf
ANTHOS: (G) flower
DEBILIS: (L) weak, feeble

Flowers in some species of this genus of over 700 species are borne on leaflike dilated branches. Weak and feeble are words that could be applied to this small annual weed.

Physalis heterophylla Nees Clammy Groundcherry

(Fiss-ah-liss het-er-oph′-ih-la)

PHYSALIS: (G) bladder; bubble
PHYSA: (G) bladder; bellows
HETERO: (G) other, different
PHYLLA: (G) leaves

The generic name is derived from the inflated calyx of this and some other species. The inflated bladderlike pod resembles a Chinese lantern. Alternate leaves of this perennial species vary from broadly ovate to rounded or heart shaped and give credence to the specific name.

Pistia stratoites L. Water lettuce

(Pis′-t-uh stra-toy′-teez)

PISTOS: (G) pure, authentic, genuine
PISTIS: (G) confidence, faith
PISTOS: (G) liquid or watery
STRATA: (L) pl. of *stratum* = a layer
STERNERE: (L) to strew, spread out; to lay flat
ITES: (L) suffix meaning belonging or related to

The generic and specific names are both appropriate to behavior rather than appearance. It is a floating, pure, authentic, obligate aquatic herb whose leaves cluster on very short branches at nodes of rootstocks. Leaves grow up to 25 cm long, are dilated upward, and result in a flat, floating layer of plants.

Plantago major L. Broadleaf Plantain

(Plan′-tay-go may′-jor)

PLANTA: (L) literally sole of the foot; a flat spreading shoot
PLANTO: (L) to plant, set
MAJOR: (L) greater

Plantar warts occur on the soles of feet. This common lawn weed does not resemble a wart, nor does it cause them, but it does spread out its broad, flat, prominently three to five ribbed leaves to cover a large area. Its larger leaves make it the major species.

Poa annua L. Annual Bluegrass

(Po′-ah an′-u-ah)

POA: (G) grass or fodder
POE: (G) grass, a grassy place
ANNUS: (L) annual, yearly

This species is an annual member of the bluegrass family. The generic name *Poa* means grass or fodder, but it is unlikely that this grass was ever used for fodder because it is so small. Other members of the genus, however, are fodder grasses. There is no question about its aggressive ability to invade turf grass, where in spite of its annual life cycle it seems perennially present.

Polygonum aviculare L. Prostrate Knotweed

(Po-lig′-o-num a-vick′-u-lar)

POLYS: (G) much, many
GONOS: (G) offspring; seed
GONATOS: (G) the knee
GONIA: (G) a corner; knee; joint; an angle
AVICULARIA: (L) small bird
AVICULA: (L) pl. *aves* = a bird
ARIA: (L) suffix meaning like or connected with

Polygonum convolvulus L. Wild Buckwheat

(Po-lig′-o-num con-vol′-view-lus)

POLYGONUM: see above
CONVOLVERE: (L) to roll together, roll up
CON: (L) together
VOLVERE: (L) to roll
VOLVO: (L) to roll

Polygonum pensylvanicum L. Pennsylvania Smartweed

(Po-lig′-o-num pen-sill-vay′-nih-cum)

POLYGONUM: see above
PENNSYLVAN: (L) fr. Pennsylvania
ICUM: (L) used as a suffix with a place name to mean of or from

Polygonum means many knees and refers to the prominent nodes on the stems of this annual weed. Each node is sheathed with a papery stipule, which aids identification of members of the genus. The reference to a small bird offered by the specific name *aviculare* is not clearly related to any conspicuous feature, unless it is the winglike papery stipule.

The specific name *convolvulus* makes sense in terms of plant behavior. It is an annual bindweed with trailing, twining stems that can grow up to 3 m long, and like perennial field bindweed *Convolvulus arvensis,* it can roll a crop together. It has small flowers in axillary clusters or on short racemes. It is easily distinguishable from field bindweed by this difference and by differences between the annual root of this species and the perennial root of *C. arvensis.*

A native of Pennsylvania, *P. pensylvanicum* was given its name by Linnaeus who did not travel to Pennsylvania but had colleagues who did and returned specimens to him. It is common in many other areas of the United States and, indeed, may not be a native of Pennsylvania.

Portulaca oleracea L. Common Purslane

(Por-chu-la′-uh o-ler-a′-c-uh)

PORTO: (L) to carry
LAC: (L) milk
OLERACEUS: (L) potherb
OLEROS: (G) impure, turbid
ACEUS: (L) suffix indicating resemblance

P. oleracea is a succulent annual with red stems and thick, somewhat wedge-shaped leaves. It does not contain a milky juice but is very succulent (water content over 90%) and the etymology fits. The specific name was assigned because purslane has been and still is eaten. Thus, the Latin derivation fits and the Greek does not. Young shoots can be served fresh in a salad or boiled as a potherb (like spinach).

Potamogeton nodosus Poir. American Pondweed

(Po-ta-mo-gee′-ton no-doe′-sus)

POTAMOS: (G) a river

POTOS: (G) a drinking
POTES: (G) a drinker
GEITON: (G) a neighbor
NODULUS: (L) knotted, knobby
NODOSUS: (L) full of knots

All potamogetons are annual or perennial herbs that propagate by seeds. They are neighbors or residents of rivers and very common in ponds and fresh or slightly brackish water. They do not grow well in swiftly running, deep water but occur as neighbors on river and creek banks. The specific name was probably assigned because of the knobby or rough nature of the seed coat.

***Pteridium aquilinum* (L.)Kuhn** Bracken Fern
(Ter-id'-e-um ack-wih-line'-um)

PTERON: (G) winged, wing
PTERIDION: (L) wing, fin
IUM: (L) dim. suffix
AQUILINE: (L) prominent
AQUILA: (L) eagle
UM: (L) suffix indicating possession

The leaves of this perennial are like fern fronds. They have a long, slender, straw-colored petiole, and the leaf blade is more or less triangular in outline and resembles a wing. Leaf blades are once to thrice divided and deeply lobed below the apex. We often see reference to the aquiline curve of someone's nose. It is usually a large, prominent nose that curves downward, like an eagle's beak, when viewed in profile. The ends of fronds of this fern curve backward on themselves and could resemble the aquiline curve of an eagle's beak. Leaf edges also curve downward and form a protective covering for spores; either could be the source of the specific name.

***Pueraria lobata* (Willd.)Ohwi** Kudzu
(Pew-rare'-e-uh lo'-bah-ta)

PUERARIA: after Marc Nicola Puerari, a Swiss botanist, died 1845
LOBATA: (L) lobed
LOBATUS: (L) lobed
LOBUS: (L) lobe

This high-climbing, perennial, leguminous vine has entire or coarsely palmately lobed leaves, which gives it the specific name.

Quercus geminata Small Scrub Live Oak
(Quare′-cus gem-in-nah′-ta)

QUERCUS: (L) oak
QUER: (C) fine
CUEZ: (C) a tree
GEMINATA: (L) double, paired

This genus of about 500 species consists of shrubs to large trees, many of which are fine oak trees. The oaks in the genus range from large oak trees in forests to smaller scrub oaks that are weeds to cattlemen who fight to keep them off rangeland. The double or paired feature of Q. *geminata* is found in its acorns, which are commonly borne in pairs at the base of leaves.

Rhus radicans L. (syn. *R. toxicodendron* L.)
Poison Ivy
(roos rad′ih-cans) (tox-ih-co-den′-dron)

RHUS: (L) sumac
RHOUS: (G) sumac
RADIX: (L) root, of or relating to roots
TOXIKOS: (G) poison
DENDRON: (G) *drys* = tree

Poison ivy belongs to the sumac family, the Anacardiaceae, and is related to poison sumac (*Rhus vernix* L.) and poison oak (*Rhus diversiloba* T. & G.). Sometimes it has an aerial root and subterranean stolons, which may be the origin of the specific name. It is likely that one will also encounter the name *Rhus toxicodendron* L. for poison ivy. The genus is very confusing and that may be because taxonomists do not like to work with something that makes them itch. In any case, the specific name *toxicodendron* comes from the Greek *dendron* (akin to *drys* = tree) and means that its roots are toxic.

Rottboellia exaltata L.f. (syn. *R. cochinchinensis* [Lour.]W.D. Clayton) Itchgrass
(Rot-bowl′-e-uh x-all-tay′-ta) (co-chin-chin-n′-sis)

ROTTBOELLIA: after Christen F. Rottboell, a Danish botanist, died 1797
EXALTARE: (L) to make high, raise, or lift up

COCHIN: (L) may refer to origin in Cochin, a city in southwestern India
CHIN: (L) may refer to a second site of origin in China
ENSIS: (L) suffix meaning place of origin

A native of India, this annual grass is a very serious weed of upland tropical crops and crops in warmer temperate climates. It is difficult to weed by hand because leaf sheaths and blades are covered with short hairs that break off on contact, are irritating, and easily penetrate one's skin. The grass grows 1–3 m high and can easily exalt itself over its neighbors. The newer specific name *cochinchinensis* agrees and disagrees with the Indian origin.

Rudbeckia hirta **var.** ***pulcherrima*** **Farw.** Black-eyed Susan

(Rud-beck'-e-uh her'-ta) (pul-chair'-e-ma)

RUDBECKIA: after Olaf Rudbeck, a Swedish scientist, died 1702
HIRTUS: (L) rough, hairy
PULCHER: (L) beautiful
RIMA: (L) slit, fissure, or crack

The erect, simple stems and leaves are rough and hairy throughout and give plants a rough feel. The varietal name *pulcherrima* must refer to the pretty, central purple-brown disk flowers and surrounding yellow ray flowers. The varietal name suffix may be for euphonius reasons without relation to appearance.

Rumex crispus **L.** Curly Dock

(Roo'-mex cris'-pus)

RUMEX: (L) sorrel, light red-brown color
RUMICIS: (L) sorrel
CRISPUS: (L) curled, wrinkled
CRISPO: (L) to curl

Plants of this genus used to be called sorrel, and the inflorescence is a light brown or sorrel color at maturity. The simple, lanceolate leaves have prominent curly or wrinkled margins; thus, the scientific name is appropriate.

Saccharum spontaneum L. Wild Sugarcane
(Sack′-r-um spon-tan′-e-um)

SAKCHAR: (G) *sakcharon* = sugar
SINGKARA: sugarcane
ARUM: (L) fr. *arundin* = reed
SPONTANEUS: (L) of free will
SPONTE: (L) free will, voluntarily

Wild cane, like domesticated cane, has sweet juice, but wild cane apparently does not produce as much sugar and is a weed in sugarcane fields. It also is a volunteer and arises of its own free will as any good weed should.

Salix humulis Marsh. Prairie Willow
(Say′-lix hume′-u-lis)

SALIX: (L) willow
SALICIS: (L) the willow
HUMULIS: (L) low growing; small
HUMUS: (L) the ground, soil
HUMI: (L) on the ground

Willows are prostrate or ascending shrubs to large trees, and *Salix humulis* is a willow that grows along the ground. This species grows only 1–3 m high.

Salsola iberica Sennen & Pau Russian Thistle
(Sal′-so-la i-beer′-ih-ca)

SALSUS: (L) salted
IBERICA: (L) fr. Spain, Spanish origin

The generic name comes from the ability, perhaps even the preference, of this weed to grow on saline sites or from the abundant alkali yielded by its ashes. It also is apparently of Spanish origin.

Salvinia auriculata Aubl. Giant Salvinia
(Sal-vin′-e-uh o-ric-u-la′-ta)

SALVINIA: after Antonio M. Salvini, an Italian linguist, died 1729
AURICUL: (L) fr. *auricula*, dim. of *auris* = ear

ATA: (L) suffix indicating possession or likeness
AURICULATA: (L) furnished with auricles

Salvinia molesta Mitch. Karibaweed

(Sal-vin′-e-uh mo-les′-ta)

SALVINIA: see above
MOLESTARE: (L) to burden; to annoy
MOLESTUS: (L) disturbed

This small genus includes minute aquatic plants that are free floating or mud rooted with few roots, have leaves modified as roots, or have a rhizome. The specific name *auriculata* was assigned because of numerous, cagelike, club-shaped hairs on the upper surface of aerial leaves and their resemblance to the shape (not position) of an auricle.

S. molesta is a particularly burdensome, disturbing weed because it has invaded large areas of water and is hard to control. The name was assigned by D. S. Mitchell, an Australian, who has studied its weedy behavior.

Saponaria officinalis L. Bouncing Bet

(Sap-o-nare′-e-uh o-fis-in-al′-iss)

SAPONIS: (L) soap
SAPO: (L) soap
ARIA: (L) a suffix meaning belonging to or connected with
OFFICINA: (L) an office
OPIFICINA: (L) originally a storeroom or workshop, later pharmacy
ALIS: (L) suffix meaning pertaining to

Saponaria got its name because some species produce a lather when leaves are crushed and rubbed by hand. The specific name *officinalis* tells us that at one time it was used for medicinal purposes. The name comes from the Latin for storeroom and later pharmacy, where medicinal plants were kept and dispensed. One frequently finds this specific name associated with a plant, but the exact use has been lost or neglected.

Scirpus mucronatus L. Ricefield Bulrush

(Scur′-pus mew-cro′-na-tus)

SCIRPUS: (L) rush, bulrush
MUCRO: (L) point; edge
ATUS: (L) suffix meaning provided with

Scirpus means rush and this genus includes annual and perennial, usually aquatic, rushes. Leaves of this species end in sharp points making the specific name descriptive.

***Senecio vulgaris* L.** Common Groundsel

(Sen-e′-c-o vul-gare′-iss)

SENEX: (L) old or old man
SENICULUS: (L) an old man
SENESCO: (L) to grow old
VULGATUS: (L) common, ordinary
VULGARE: (L) to make known
VULGUS: (L) mob, common people

Senecio is Latin for old or old man. Such men often have white hair and the name may have been assigned because of the hoariness of many species or because of white hairs of the pappus surrounding the achene. The specific name *vulgaris* means it was common in many sites. It may also mean that the plant was commonly found on sites associated with humans so it was seen frequently even though it did not grow everywhere.

***Sesbania exaltata* (Raf.)Rydb. ex A.W.Hill** Hemp

(Sez-bane′-e-uh x-all-ta′-ta)

SESBANIA: (Per) *sisaban* = rope, fiber; kind of tree
EXALTO: (L) to make high, raise, lift up
ATA: (L) suffix indicating possession or likeness

The generic name may be of Persian origin or it may be a Latinized version of *Sesban Adans,* which is probably of Arabic origin. Some species produce a tough, woody fiber used to make ropes, matting, or fabrics. *S. exaltata* grows up to 3 m tall when undisturbed.

***Setaria faberi* Herrm.** Giant Foxtail

(Seh-tare′-e-uh fa-ber′-i)

SETA: (L) bristle, having bristles
ARIA: (L) suffix meaning like or connected with
FABERI: (L) after Ernst Faber, who named the species in 1910

Setaria glauca (L.)Beauv. Yellow Foxtail
(Seh-tare′-e-uh glaw-ca)

SETARIA: see above
GLAUCA: (G) fr. *glaukos* = silvery; gleaming

Setaria verticillata (L.)Beauv. Bristly Foxtail
(Seh-tare′-e-uh ver-tih-sih′-la-ta)

SETARIA: see above
VERTICILA: (L) whorled
ATA: (L) suffix indicating possession or likeness

Setaria viridis (L.)Beauv. Green Foxtail
(Seh-tare′-e-uh veer′-ih-dis)

SETARIA: see above
VIRIDIS: (L) green

Setaria means possessing bristles or short hairs. The inflorescences of all species of this genus possess hairs in abundance and give the plants their common name, foxtail.

S. glauca resembles S. verticillata, which has a more spreading and branching habit. *S. verticillata* also has rougher stems and leaves and usually only one downwardly barbed bristle below each spikelet. The relative paucity of bristles may give it an appearance of having whorls in the inflorescence, but there are none, and the name is not appropriate.

The inflorescence of *S. viridis* is dominantly green, but this is not a positive identifying trait.

Sida acuta Burm.f. Southern Sida
(Side′-uh ah-q′-ta)

SIDE: (G) a water plant or a pomegranate tree
SIDA: (G) a nymph
ACUTA: (L) sharp or acutely angled

Sida spinosa L. Prickly Sida
(Side′-uh spin-o′-sa)

SIDA: see above
SPINOSUS: (L) full of thorns
SPINULA: (L) thorn or spine

The etymology of this genus is confusing. *Side* in Greek means water plant, but these are all terrestrial plants and not succulent. The name was used by Theophrastus for unknown reasons. Leaves of *S. acuta* are lanceolate and thus have a pointed end that could be regarded as acute angled. They also have serrated margins with acute angles.

The etymology of the specific name *spinosa* is not immediately obvious because the plant has softly pubescent leaves and stem. The spines, which may have been the basis for the name, are hardened leaf stipules at leaf bases.

***Silene alba* (Mill.)E.H.L.Krause** (syn. *Lychnis alba* Mill.) White Campion

(Sigh′-leen al′-ba) (Lich′-nis)

SILENE: (G) prob. fr. Silenos, the foster father and companion of Dionysus
SIALON: (G) spittle, foam
SILERE: (L) to be silent
ALBA: (L) white

The generic name probably comes from Silenos or Silenus the foster father and companion of Dionysus. He took a human form but had the ears and tail of a horse. Occasionally, he appears with the legs of a horse or goat. He is frequently drunk, always old, frequently bald, always bearded, and often covered with foam. He led the satyrs and was a constant attendant of Bacchus. The relationship of this etymology to the over 500 species of the genus is not clear. The generic name could also be derived from the Latin *silere,* to be silent. Some species bloom only at night when it is usually silent. Finally, the name could be from the Greek *sialin* or *salvia,* which was a name given to many plants producing viscid moisture on their stalk. This may be the proper origin and derivation of the common name catchfly for some weedy species. We do not know for sure. This plant has five-petalled white flowers. See also *Lychnis alba.*

***Sinapis arvensis* L.** Wild Mustard

(Sin-ap′-is r-ven′-sis)

SINAPIS: (L) mustard
ARVENS: (L) of the field
ARVUM: (L) arable field

The etymology of this plant name is easy and appropriate. This is a field mustard and has not been cultivated by humans. The name probably comes from the resemblance of the yellow flowers to the mustard color.

Solanum nigrum **L.** Black Nightshade

(So-lay′-num ni′-grum)

SOLA: (L) sun
SOLOR: (L) to comfort, soothe
SOLARI: (L) to comfort or quiet
SOLAMEN: (L) a solace
SOLANUM: (L) literally means nightshade
NUM: (L) an interrogative particle usually implying a negative answer
NIGRA: (L) black
UM: (L) a suffix indicating possession

Solanum was probably assigned because deadly nightshade *Atropa bella-donna* was one of the first plants classified in this genus. Atropine is extracted from it, and legend tells us that witches brewed the potent plant juice in the dead of night. The name could also have been derived from *solamen,* Latin for solace or comfort, because of medicinal and sedative properties of these plants. *Nigrum* refers to the black color of ripe fruit.

Solidago canadensis **L.** Canada Goldenrod

(Sol-ih-day′-go can-uh-den′-sis)

SOLDARE: (L) to make whole or sound
SOLIDO: (L) to put together
SOLDAGO: (L) an herb reputed to heal wounds
CANAD: (L) fr. Canada
ENSIS: (L) suffix meaning place of origin

The generic name alludes to the vulnerary qualities of these plants. It tells us that when they were used for medicinal purposes they could make one whole again or cure open wounds. This species is a native of Canada.

Sonchus arvensis **L.** Perennial Sowthistle

(Sahn′-cus r-ven′-sis)

SONCHOS: (G) sowthistle
SOOS: (G) safe
ECHEIN: (G) to have
ARVENS: (L) of the field
ARVUM: (L) an arable field

Sonchus oleraceus L. Annual Sowthistle
(Sahn'-cus o-ler-a'-c-us)

SONCHUS: see above
OLEROS: (G) impure, turbid
OLUS: (L) garden, potherb
OLERIS: (L) garden, potherb
HOLUS: (L) garden, potherb
ACEUS: (L) likeness, resemblance

Sonchos is Greek for sowthistle. There is a possibility that the name could have been derived from a combination of Greek *soos,* safe, and *echein,* to have, because of the perception of healthful juice derived from some species. All members of this genus are leafy-stemmed, coarse, succulent weeds with a milky latex. It is not clear which species this may have applied to. *S. arvensis* is a perennial weed of many fields in the temperate zones.

S. oleraceus is an erect annual herb grows 0.3 to 1.3 m tall. When young, it can be used as a potherb, but older plants quickly lose their tenderness and become coarse and inedible. The milky (perhaps turbid) latex imparts a bitter taste and may have been regarded as impure.

Sorghum halepense (L.)Pers. Johnsongrass
(Sor'-gum hal'-eh-pence)

SYRICUS: (L) fr. Syria
SURGUM: (L) great millet
GRANUM: (L) grain
HALEPOS: (G) barbed, severe
HALEPENSIS: (G) literally from Aleppo, Syria

The etymology of *Sorghum* is not clear, but it is close to a combination of the Latin *syricus* or Syria, and *granum,* grain. Syria is the proposed place of origin for this weed. The proposal is further confirmed by the specific name, which literally means from Aleppo, a city in Syria. Modern sorghum varieties are sources of grain for people and cattle and there is no doubt that relatives of this weed were used as grain.

Sparganium eurycarpum Engelm. Giant Burreed
(Spar-gan'-e-um u-ree-car'-pum)

SPARGANION: (L) burreed

SPARGANON: (G) swaddling band; ribbon
EURY: (G) broad
URU: (San) broad
KARPOS: (G) fruit

The generic name is probably derived from a resemblance of the ribbonlike or linear leaves to a band or linear swaddling board. Fruit of this species has a broad, flattened top with a shallow notch as opposed to the more linear fruit of other species.

Spergula arvensis L. Corn Spurry

(Sper-goo′-la r-ven′sis)

SPERGERE: (L) to scatter, to strew
SPARGO: (L) I scatter
ARVENS: (L) of the field
ARVUM: (L) an arable field

Some members of this genus are predominantly spreading weeds and do not grow tall. They elongate and can thus scatter or spread out to disperse seed.

Sphenoclea zeylandica Gaertn. Gooseweed

(Spheh-noc′-lee-uh zay-lan′-dih-cah)

SPHENOS: (G) wedge
SPHENARION: (G) wedge shaped; spoon
SPHENO: (G) comb. form meaning wedge shaped
CLEA: (L) to enclose
KLEIO: (G) to enclose, shut up
ZEYLANICA: (L) fr. Ceylon (Sri Lanka)

Leaves of this species do not match the generic description of being spoon or wedge shaped because they are oblong to lanceolate and narrow to a point at the tip. However, the floral bracts are only 2–3 mm long and about 1 mm wide but are definitely spatulate or spoon shaped thus giving this monotypic genus its name. It may be a native of Ceylon (Sri Lanka) and is definitely an Old World species.

Sporobolus cryptandrus (Torr.)Gray Sand Dropseed
(Spore-ob′-o-lus cryp-tan′-drus)

SPORA: (G) seed
BOLOS: (G) throw
BALLEIN: (G) to throw
SPORADOS: (G) scattered
KRYPTOS: (G) secret, hidden
ANDRUS: (G) comb. form meaning male

The generic name is inaccurate because the plant does not throw its seed as some plants do. The palea splits at maturity and grain falls to the ground, giving the common name dropseed, which is more accurate than the reference to throwing seed. One-flowered spikelets of this species do not have prominent emerging anthers, as many grass species do, to facilitate pollination.

Stellaria media (L.)Vill. Common Chickweed
(Stell-r′-e-uh me′-d-uh)

STELLA: (L) a star
STELLATE: (L) divided into segments; radiating from a common center
MEDIA: (L) medium
MEDIUS: (L) middle

The flowers of this annual are perfect and occur in leaf axils or in cymose clusters. There are five sepals and five two-parted petals resembling a star, which gives the plant its generic name. The species is called *media* because it is a medium-sized member of the genus.

Stipa comata Trin. & Rupr. Needle-and-Thread
(Sti′-pa co′-ma-ta)

STIPA: (L) the coarse part of flax
COMA: (L) hair
ATUS: (L) suffix meaning provided with

The generic name was traced to the Latin *stipa,* which means the coarse part of flax. This is an allusion to the flaxen appearance of feathery awns of the original species. There is no resemblance of this species or of other members of the genus to the gross appearance of flax. The reference of the specific name *comata* to hair is easy to associate with this perennial grass if one's hair is unruly and wind blown. One-flowered spikelets occur in a loose, erect, spreading panicle. Empty glumes taper into a slender awn, which is loosely twisted and indistinctly twice bent or angled. It is often deciduous but also serves as a way to spread seed when it becomes attached to cloth or fur.

***Striga asiatica* (L.)Ktze.** Witchweed

(Stri′-ga a-z-at′-ih-ca)

STRIGA: (L) furrow, windrow, or swath; also a straight, rigid, close-pressed, short bristlelike hair
ASIATICA: (L) fr. Asia

***Striga hermonthica* (Del.)Benth.** Witchweed

(Stri′-ga her-mon′-thih-ca)

STRIGA: see above
HERMON: (G) a statue in the form of a square stone or pillar surmounted by a bust or head
HERMES: (G) the god who served as herald and messenger for other gods
HICA: (L) fr. *hice* = like

The *Striga* genus contains about 25 species, a few of which are very important parasitic weeds on a wide range of crops, especially cereal crops. The generic name was given because of calyx ribs or grooves. All species have at least five grooves. This one has ten or more but never more than five extend to the tip of the calyx lobes. *S. asiatica* is native to Asia and is now found in a few counties in North and South Carolina in the United States, where it is tightly quarantined; the U.S. aim is eradication. It is a dominant weed of cereals in many of the tropical and semitropical areas of the world.

The calyx of *S. hermonthica* is distinctly five ribbed with a bright pink corolla tube beginning at and apparently perched on the calyx pedestal, making it appear like a pedestal with a figure on it. There is no evident relationship to Hermes. The corolla tube is 11–17 mm long and bends at an angle immediately over the calyx tip.

***Tanacetum vulgare* L.** Common Tansy

(Tan-ah′-c-tum vul′-gare)

TANACETUM: (G) prob. fr. *anthanasia* = immortality
TENACETUM: (L) a name for tansy
VULGATUS: (L) ordinary, common
VULGARE: (L) to make known
VULGUS: (L) mob, commom people

The generic name comes from the Greek *anthanasia,* or immortality, a reference to its medicinal qualities. Foliage of members of this genus has been used to preserve corpses, preserve meat in summer, repel insects, and for various medicinal purposes, for which it was imported to the United States from Europe. It was once regarded as a specific medicant for intestinal worms and was used in New England and Europe in funeral winding sheets to discourage worms from entering the body. This species is a common or ordinary perennial weed often found where humans have been and reproduces by seeds and creeping rootstocks.

***Taraxacum officinale* Weber in Wiggers** Dandelion

(Tare-ax′-ah-cum o-fis-in-al′)

TARASSE: (G) to disturb; to disorder
TARAXIS: (G) disorder, confusion
TARAKHSHAQUN: (A) wild chicory
TALKH CHAKOK: (Per) bitter herb
OPIFICINA: (L) originally a storeroom or workshop, later pharmacy
OFFICINA: (L) an office

The generic name may come from any one or all of three sources: Greek, Arabic, or Persian. The specific name comes from two Latin words for office or storeroom or, later, pharmacy. Use of a term with these roots implies medicinal value and that the plant may have been included on the official (*officinale*) drug list and kept in stock by druggists. Therefore, we have a bitter herb similar to wild chicory (see *Cichorium intybus*), which disturbs us, but also one that at one time was useful and, in fact, extracts of which may have been dispensed by pharmacists. Dandelion leaves are used to make wine and for many other food purposes.

***Torenia concolor* Lindl.** Torenia

(Tor-e′-nee-uh con′-color)

TORENIA: after Olaf Toren, a ship's chaplain, died 1753

CON: (L) *cum* = with
COLOR: (L) color

No significance can be attached to the fact that this plant bears purple flowers; many do. We also do not know why the genus was named for a ship's chaplain. This is a good example of an apparently meaningless name.

Tragopogon pratensis L. Meadow Salsify

(Tray'-go-po-gon pray-ten'-sis)

TRAGOS: (G) goat
TRAGULUS: (G) a male goat
POGON: (G) beard
PRATUM: (L) a meadow
PRATENSIS: (L) growing in fields or meadows

The achenes of *T. pratensis* are 10–15 mm long and terminate in a long beak. Each seed of this species and all other species of the genus has a pappus of numerous yellow, plumose bristles. When the pappus is collected together, as it is on the plant, by an involucre of long, pointed, narrow bracts that tend to constrict it, it resembles a goat's beard, thus making the scientific name appropriate. Meadow salsify grows in meadows but the etymology of salsify is not clear. It is probably from the Latin *saxifrica,* which denotes any of various herbs, usually with purple flowers.

Trianthema portulacastrum L. Horse Purslane

(Tri-an'-theme-ah por-chu-la-cas'-trum)

TRI: (G) three
ANTHEMA: (G) flower
ANTHOS: (G) a comb. form meaning flower
PORTULA: (L) a small door or gate
PORTA: (L) door, gate
CASTRUM: (L) any fortified place

Some members of this small genus may have three anthers or three flowers as the name implies; this one does not have either. However, it does have flowers that are sessile and usually solitary in leaf axils. These flowers and the underlying perigynous ovary are partly hidden because the ovary is embedded in and thus hidden by the receptacle and concealed by the petiolar sheath.

Tribulus cistoides L. Jamaica Feverplant
(Trib′-u′-lus sis-toy′-deez)

TRIBOLOS: (G) any of various prickly plants; three pointed
CISTA: (L) basket, box, or chest
KISTA: (G) box or chest
OIDES: (L) suffix meaning like or resembling

Tribulus terrestris L. Puncturevine
(Trib′-u-lus ter-res′-tris)

TRIBULUS: see above
TERRA: (L) earth
TERRESTRIS: (L) of the earth or soil, particularly dry soil

The generic name comes from the characteristic seed pod. Pistils are united into five carpels and one has only a few seeds. At maturity the hard carpels separate and have two to four stiff, spreading spines each up to 7 mm long. Carpels can survive in the soil for several years awaiting conditions favorable for germination. The spines can penetrate tires and skin with equal ease. Linnaeus undoubtedly saw a resemblance of the seed pod to the Roman caltrop, a four-pointed, metallic device that rested on three prongs and had a fourth pointing straight up to act as a deterrent to cavalry and foot soldiers. The specific name *cistoides* also refers to the seed pod, which when opened resembles a small box or chest complete with a hollow cavity for seeds.

T. terrestris is the most common weedy species found in farm fields. Its seed pods are commonly found in flat bicycle tires, especially in more arid areas and other favored habitats.

Tridax procumbens L. Coat Buttons
(Tri′-dax pro′- cum-benz)

TRI: (L) three
DAX: (G) prob. fr. *dakytlos* = finger
PRO: (L) forward
CUMBERE: (L) to lie down
PROCUMBERE: (L) to fall, bend, or lean forward

The generic name was derived from the three lobes of the white ray florets of this short-lived perennial or annual weed. The specific name was probably derived from the fact that the flower heads bend forward on their slender, erect stalks. It may also have been derived from the decumbent habit of stems, which can root at their nodes.

Tripleurospermum maritimum (L.)W.D.J.Koch
(syn. *Matricaria maritima* L.) False Chamomile
(Triple-u-ro-sper′-mum mare-ih-t′-mum) (Mat-trih-care′-e-uh mare-ih-t′-ma)

TRI: (L) three
PLEURO: (G) lateral; ribbed in a sideways position
SPERMUM: (L) *spermus* = seeded
SPEIREIN: (G) to sow
SPERMA: (G) seed
MARITIMUS: (L) growing by the sea

The old generic name *Tripleurospermum* is one that cannot be resisted when exploring etymology. It is euphonius and accurate. *T. maritimum* bears a three-ribbed seed. The specific name is not quite as reliable. Surely it can grow by the sea, but it is not bound to do so and is common on many agricultural sites. Some members of the genus *Matricaria* were used in folk medicine for treatment of menstrual disorders and this gives the now accepted generic name validity. See also *Matricaria maritima*.

Tussilago farfara L. Coltsfoot
(Too′-sih-la-go far-far'ah)

TUSSILAGO: (L) coltsfoot
TUSSIS: (L) a cough
FARFARA: (S/P) coltsfoot

The scientific name has been traced to its common name, coltsfoot, and no farther. We know *tussis* means a cough in Latin, and this may be a more meaningful etymology than coltsfoot because it reveals a now unknown medicinal use. Another common name is cough-wort, which implies use for medicinal properties, probably cough medicine.

Typha latifolia L. Common Cattail
(Tie′-fa lah-tih-fo′-lee-uh)

TYPHE: (G) a cat's tail
TYPHOS: (G) smoke or cloud
LATUS: (L) broad, wide
FOLIA: (L) leaves

There are about 10 species in this genus of widely distributed marsh plants, and they cause problems in irrigation canals because they impede water flow. They also cause problems in shallow shore areas of lakes and ponds used for recreation and fishing. Staminate flowers are grouped in a light yellow spike at the top of the stem; pistillate flowers are grouped below. After pollen has been shed, the staminate flowers fall and only a naked spike remains. The pistillate flowers form the round, cigar-shaped cat's tail, which can resemble smoke or a cloud when small seeds enveloped in the "cotton" or hairs of the cattail are released. The long, alternate, grasslike leaves are 6–25 mm broad and the specific name describes them well.

***Urtica dioica* L.** Stinging Nettle

(Ur'-tih-ca die-o'-ih-ca)

URTICA: (L) nettle
URO: (L) to burn, to sting
URTICARE: (L) to sting
DI: (G) two
OIKOS: (G) house

This perennial grows from a rootstock and produces erect, 1- to 2-m-tall stems. The stems are ridged, bristly, and hairy. Contact with the hairs causes a stinging sensation followed, in most people, by itching and slight welts. It is not as severe or contagious as poison ivy but it can be irritating even if only for a short time. Nettles are dioecious with female flowers on one plant and male flowers on another. In early morning male flowers puff (release) their golden pollen into the air where it may be caught by female flowers of a neighboring plant. Thus, the specific name *dioica* (*di* + *oikos*), two homes, fits.

***Veratrum californicum* Durand** California False Hellebore

(Ver-at'-trum cal-ih-for'-nih-cum)

VERUS: (L) true
ATER: (L) black
ATRUM: (L) black
CALIFORN: (L) fr. California
ICUM: (L) used as suffix with place name to mean of or from

This stout, tall, leafy perennial is not truly black. It has broad, sessile, generally oval, coarsely veined leaves that sheath the plant at its base. Each is 15–30 cm long and all are dark green. Flowers are in a panicle 30–45 cm long. Each flower is dull white and the perianth segments may have a green margin. The generic name was probably derived from the fact that these plants are poisonous to livestock and they smell bad, but there is no clear explanation for their being called truly black.

Verbascum thapsus L. Common Mullein

(Ver-bas′-cum thap′-sus)

VERBASCUM: (L) mullein
THAPSOS: (L) fr. Thapsos, Sicily
THAPSUS: (G) fr. Thapsus, Tunisia

Verbascum is the name given by Pliny. It is probably a corruption of the Latin *Barbascum,* which meant a hairy plant and came from the Latin *barba,* or beard. *V. thapsus* could have originated in Thapsos, Sicily, or Thapsus, Tunisia, as both are logical and fit the specific name.

Vernonia cinerea (L.)Less. Little Ironweed

(Ver-no′-nee-uh sin-′-er-e-uh)

VERNONIA: after William Vernon, an English botanist, died 1711
CINERARIUS: (L) pertaining to ashes
CINERIS: (L) ashes
CINERE: (L) ash colored

This erect annual herb is sparingly branched and slightly pubescent. The specific name comes from the pale purple or blue color of terminal flowers, colors resembling ashes.

Veronica agrestis L. Field Speedwell

(Ver-on′-ih-ca ah-gres′-tis)

BERENIKON: (L/G) a plant
VERONICA: (L) a comb. form of *vera* and *icon* = a true image
AGRESTIS: (L) rural
AGER: (L) field
AGROS: (G) country

This common annual has a generic name composed of two words from a late medieval legend that mean true image. In the legend a maiden handed her handkerchief to Jesus on his way to Calvary. He wiped sweat from his brow and returned the handkerchief, which bore a perfect likeness of his face—a *vera icon,* or true image. The maiden became St. Veronica and is commemorated on February 4th.

Vicia sepium L. Hedge Vetch
(Vih′-c-uh c′-p-um)

VICIA: (L) vetch
VINCERE: (L) to bind
VINCIO: (L) to bind together
SEPIUM: (L) pl. *sepes* = a hedge or enclosure

Vicia may come from the Latin *vincere,* which means to bind and is an allusion to the dense foliar mat it forms. It is not a hedge-forming plant, but the foliage is as dense as a hedge. The stems are very slender and leaves are pinnately compound with about seven pairs of leaflets and a terminal tendril by which plants climb or form mats.

Xanthium spinosum L. Spiny Cocklebur
(Zan′-thee-um spih-no′-sum)

XANTH: (G) yellow
XANTHOS: (G) the various shades of yellow
SPINA: (L) thorn or spine
SPINOSUM: (L) with spines

Xanthium strumarium L. Common Cocklebur
(Zan′-thee-um strew-mare′-e-um)

XANTHIUM: see above
STRUMA: (L) cushionlike swelling or tumor, a scrofulous tumor
STRUMOSUS: (L) swollen; a swelling of glands

Xanthium comes from the Greek word for yellow, and extracts from members of this genus were once used as cloth dye and perhaps even as hair dye. The specific name *spinosum* was assigned because the plant is armed with a pair of three-parted, sharp, yellow stipular spines, each about 2.5 cm long in axils of the petiole and main stem.

X. strumarium, an immigrant from the Old World, is an annual weed in many important crops. It is a coarse, stout plant growing 60–120 cm tall. The stems and leaves are rough to touch and the stems are spotted with irregularly shaped red-brown areas. It is not precisely clear but it seems that the specific name comes from the fruiting bur (the "cocklebur") that is densely set with hooked yellow prickles, each 3–7 mm long. These are often glandular and sparsely pubescent at the base. The fruiting bur in this species is much more prickly than in *X. spinosum* and could be viewed as a scrofulous swelling or tumor. It is a two-seeded fruiting structure and serves, because of its spines, as a primary means of seed spread.

Yucca glauca Nutt. ex Fraser Great Plains Yucca
(Yuck-ah glaw'-ca)

YUCCA: (S) unknown origin
GLAUCUS: (L) gleaming, gray
GLAUKOS: (G) gleaming; silvery, gray, or blue-gray

The generic name is a Spanish word of unknown origin. The specific name refers to the gray-white, inverted, bell-like flowers.

Zigadenus venenosus S. Wats. Meadow Deathcamas
(Zig-ah-d'-nus ven-n-o'-sus)

ZYGADEN: (G) jointly; in pairs
ZYGON: (G) yoke; pair
ADEN: (G) a gland
VENENO: (L) to poison
VENENUM: (L) poison

The generic name is almost accurate. The plants are smooth, often glacous perennials with rhizomes or bulbs, leafy stems, and large panicles or racemes of white, yellow, or bronze-tinged flowers. Floral segments are oblong to oval and have one or two glands near their more or less narrowed base. It is these supposedly paired glands that the generic name refers to. This species is poisonous to sheep and cattle.

Zizaniopsis miliacea (Michx.)Doell & Aschers.
Giant Cutgrass
(Zih-zan-e-op'-sis mill-e-a'-c-ah)

ZIZANION: (G) wild grain
ZIZANIUM: (L) darnel
OPSIS: (G) suffix meaning appearance or resemblance
MILIACEUS: (L) consisting of millet
MILIUM: (L) millet
ACEA: (L) suffix meaning resembling or pertaining to

This perennial grass resembles annual wild rice, *Zizania aquatica,* and millet, *Panicum* sp., due to its long, narrow, terminal panicles.

Scientific Names of Weeds by Plant Family

AGAVACEAE
Yucca glauca Nutt. ex Fraser
AIZOACEAE
Mollugo pentaphylla L.
M. verticillata L.
Trianthema portulacastrum L.
AMARANTHACEAE
Alternanthera sessilis (L.) R.Br. ex DC.
Amaranthus hybridus L.
A. retroflexus L.
Celosia argentea L.
ANACARDIACEAE
Rhus radicans L (syn. *R. toxicodendron* L.)
R. toxicodendron L. (syn. *R. radicans* L.)
APIACEA
Daucus carota L.
ARACEAE
Pistia stratoites L.
ASCLEPIADACEAE
Asclepias speciosa Torr.
ASTERACEAE (Compositae)
Ageratum conyzoides L.
Ambrosia artemisiifolia L.
A. tomentosa Nutt.
A. trifida L.
Anthemis cotula L.
Arctium minus (Hill)Bernh.
Artemisia tridentata Nutt.
Bidens pilosa L.
Carduus nutans L.
Centaurea repens L.
Chromolaena odorata (L.)R.M.King & M.Robinson
Chrysothamnus nauseosus (Pallas)Britt.
Cichorium intybus L.
Cirsium arvense (L.)Scop.
C. vulgare (Savi)Tenore

Conyza canadensis (L.)Cronq.
Eclipta alba (L.)Hassk. (syn. *E. prostrata* L.)
E. prostrata (syn. *Eclipta alba* [L.]Hassk.)
Erigeron strigosus Muhl. ex Willd.
Eupatorium capillifolium (Lam.)Small
Galinosoga parviflora Cav.
Grindelia squarrosa (Pursh)Dunal
Gutierrezia sarothrae (Pursh)Britt. & Rusby
Helianthus petiolaris Nutt.
Lactuca serriola L.
Matricaria inodora L. (syn. *M. perforata* Merat)
M. maritima L. (syn. *Tripleurospermum maritimum* [L.] W.D.J. Koch)
M. matricarioides (Less)C.L.Porter
M. perforata Merat (syn. *Matricaria inodora* L.)
Mikania cordata (Burm.f.)B.L.Robins.
Rudbeckia hirta var. *pulcherrima* Farw.
Senecio vulgaris L.
Solidago canadensis L.
Sonchus arvensis L.
S. oleraceus L.
Spergula arvensis L.
Tanacetum vulgare L.
Taraxacum officinale Weber in Wiggers
Tragopogon pratensis L.
Tridax procumbens L.
Tripleurospermum maritimum (L.) W.D.J.Koch(syn. *Matricaria maritima* L.)
Tussilago farfara L.
Vernonia cinerea (L.)Less.
Xanthium spinosum L.
X. strumarium L.

BORAGINACEAE
Heliotropium indicum L.

BRASSICACEAE (Cruciferae)
Barbarea vulgaris R.Br.
Brassica kaber (DC.)L.C.Wheeler
B. nigra (L.)W.J.D.Koch

Capsella bursa-pastoris (L.)Medic.
Cardaria draba (L.)Desv.
Descurainia pinnata (Walt.)Britt.
Lepidium campestre (L.)R.Br.
Sinapis arvensis L.

CACTACEAE
Opuntia polyacantha Haw.

CAMPANULACEAE
Sphenoclea zeylandica Gaertn.

CANNABACEAE
Cannabis sativa L.

CARYOPHYLLACEAE
Agroestemma githago L.
Cerastium vulgatum L.
Lychnis alba Mill. (syn. *Silene alba* [Mill.]E.H.L.Krause)
Saponaria officinalis L.
Silene alba (Mill.)E.H.L.Krause (syn. *Lychnis alba* Mill.)
Spergula arvensis L.
Stellaria media (L.)Vill.

CERATOPHYLLACEAE
Ceratophyllum demersum L.

CHENOPODIACEAE
Atriplex canescens (Pursh)Nuttall
A. patula L.
Chenopodium album L.
Halogeton glomeratus (Stephen ex Bieb.)C.A.Mey.
Kochia scoparia (L.)Schrad.
Salsola iberica Sennen & Pau

COMMELINACEAE
Commelina benghalensis L.
C. diffusa Burm. f.
Murdannia nudiflora (L.)Brenan

CONVOLVULACEAE
Convolvulus arvensis L.
Cuscuta campestris Yuncker
Ipomoea hederacea (L.)Jacq.
I. pes-tigridis L.
I. purpurea (L.)Roth
I. triloba L.

CYPERACEAE
Cyperus difformis L.
C. esculentus L.
C. iria L.
C. rotundus L.
Fimbristylis littoralis Gaud. (syn. *F. miliacea* [L.]Vahl)
F. miliacea (L.)Vahl (syn. *Fimbristylis littoralis* Gaud.)
Scirpus mucronatus L.
DENNSTAEDTIACEAE
Pteridium aquilinum (L.)Kuhn
DIPSACACEAE
Dipsacus fullonum L.
EQUISETACEAE
Equisetum arvense L.
E. palustre L.
EUPHORBIACEAE
Euphorbia esula L.
E. hirta L.
Phyllanthus debilis Klein ex Willd.
FABACEAE
Astragalus bisulcatus (Hook.)Gray
Calopogonium muconoides Desv.
Cassia obtusifolia L.
Desmodium tortuosum (Sw.)DC.
Galega officinalis L.
Mimosa invisa Mart.
M. pudica L.
Oxytropis sericea Nutt. ex T. & G.
Pueraria lobata (Willd.)Ohwi
Sesbania exaltata (Raf.)Rydb. ex A.W.Hill
Vicia sepium L.
FAGACEAE
Quercus geminata Small
GERANIACEAE
Erodium cicutarium (L.)L'Her. ex Ait.
HYDROCHARITACEAE
Elodea canadensis L.C.Rich
Hydrilla verticillata (L.f.)Royle

HYPERICACEAE
Hypericum perforatum L.
LAMIACEAE
Lamium amplexicaule L.
LEMNACEAE
Lemna minor L.
LILIACEAE
Allium vineale L.
Veratrum californicum Durand
Zigadenus venosus S.Wats.
LYTHRACEAE
Lythrum salicaria L.
MALVACEAE
Abutilon theophrasti Medic.
Anoda cristata (L.)Schlecht.
Hibiscus trionum L.
Malva neglecta Wallr.
Sida acuta Burm. f.
S. spinosa L.
MARSILEACEAE
Marsilea minuta L.
ONAGRACEAE
Ludwigia adscendens (L.)Hara
L. octovalvis (Jacq.)Raven
OROBANCHACEAE
Orobanche crenata Forsk.
OXALIDACEAE
Oxalis corniculata L.
PAPAVERACEAE
Argemone mexicana L.
A. polyanthemos (Fedde)G.B.Ownbey
PLANTAGINACEAE
Plantago major L.
POACEAE (Gramineae)
Aegilops cylindrica Host
Agropyron repens (L.)Beauv.
Alopecurus myosuroides Huds.
Avena fatua L.

Axonopus compressus (Sw.)Beauv.
Brachiaria mutica (Forsk.)Stapf.
Bromus tectorum L.
Cenchrus echinatus L.
C. incertus M.A.Curtis
Chrysopogon aciculatus (Retz.)Trin.
Cynodon dactylon (L.)Pers.
Dactyloctenium aegyptium (L.)Willd.
Digitaria ciliaris (Retz.)Koel.
D. sanguinalis (L.)Scop.
D. scalarum (Schweinf.)Chiov.
Echinochloa colonum (L.)Link
E. crus-galli (L.)Beauv.
E. glabrescens Munro ex Hook.f.
Eleusine indica (L.)Gaertn.
Eragrostis cilianensis (All.)E.Mosher
Hordeum jubatum L.
Imperata cylindrica (L.)Beauv.
Ischaemum rugosum Salisb.
Leersia hexandra Sw.
Leptochloa chinensis (L.)Nees
Lolium multiflorum Lam.
L. temulentum L.
Panicum capillare L.
P. dichotomiflorum Michx.
P. maximum Jacq.
P. miliaceum L.
P. repens L.
Paspalum conjugatum Bergius
P. dilatatum Poir.
P. distichum Am. Auctt. (syn. *Paspalum paspaloides* [Michx.] Scribn.)
P. paspaloides (Michx.)Scribn.(syn. *Paspalum distichum* Am. Auctt.)
Pennisetum clandestinum Hochst. ex Chiov.
P. pedicellatum Trin.
P. polystachon (L.)Schultes
P. purpureum Schumach.

Phalaris arundinacea L.
Phragmites australis (Cav.)Trin. ex Steud.
Poa annua L.
Rottboellia cochinchinensis (Lour)W.D.Clayton (syn. *R. exaltata* L.f.)
R. exaltata L.f. (syn. *Rottboellia cochinchinensis* [Lour] Clayton)
Saccharum spontaneum L.
Setaria faberi Herrm.
S. glauca (L.)Beauv.
S. verticillata (L.)Beauv.
S. viridis (L.)Beauv.
Sorghum halepense (L.)Pers.
Sporobolus cryptandrus (Torr.)Gray
Stipa comata Trin. & Rupr.
Zizaniopsis miliacea (Michx.)Doell & Aschers.
POLYGONACEAE
Polygonum aviculare L.
P. convulvulus L.
P. pensylvanicum L.
Rumex crispus L.
PONTEDERIACEAE
Eichornia crassipes (Mart.)Solms
Monochoria vaginalis (Burm.f.)Kunth
PORTULACACEAE
Portulaca oleracea L.
POTAMOGETONACEAE
Potomogeton nodosus Poir.
PRIMULACEAE
Anagallis arvensis L.
Lysimachia punctata L.
RANUNCULACEAE
Delphinium geyeri Greene
RUBIACEAE
Borreria laevis (Lam.)Griseb
Galium aparine L.
Hedyotis corymbosa (L.)Lam.
SALICACEAE
Salix humulis Marsh.

SALVINIACEAE
Salvinia auriculata Aubl.
S. molesta Mitch.
SCROPHULARIACEAE
Linaria vulgaris Mill.
Lindernia anagallidea (Michx.)Pennell
L. pussilla (Willd.)Bolding
Striga asiatica (L.)Ktze.
S. hermonthica (Del.)Benth.
Torenia concolor Lindl.
Verbascum thapsus L.
Veronica agrestis L.
SOLANACEAE
Datura stramonium L.
Physalis heterophylla Nees
Solanum nigrum L.
SPARGANIACEAE
Sparganium eurycarpum Engelm.
TILIACEAE
Corchorus olitorius L.
TYPHACEAE
Typha latifolia L.
URTICACEAE
Urtica dioica L.
VERBENACEAE
Lantana camara L.
ZYGOPHYLLACEAE
Tribulus cistoides L.
T. terrestris L.

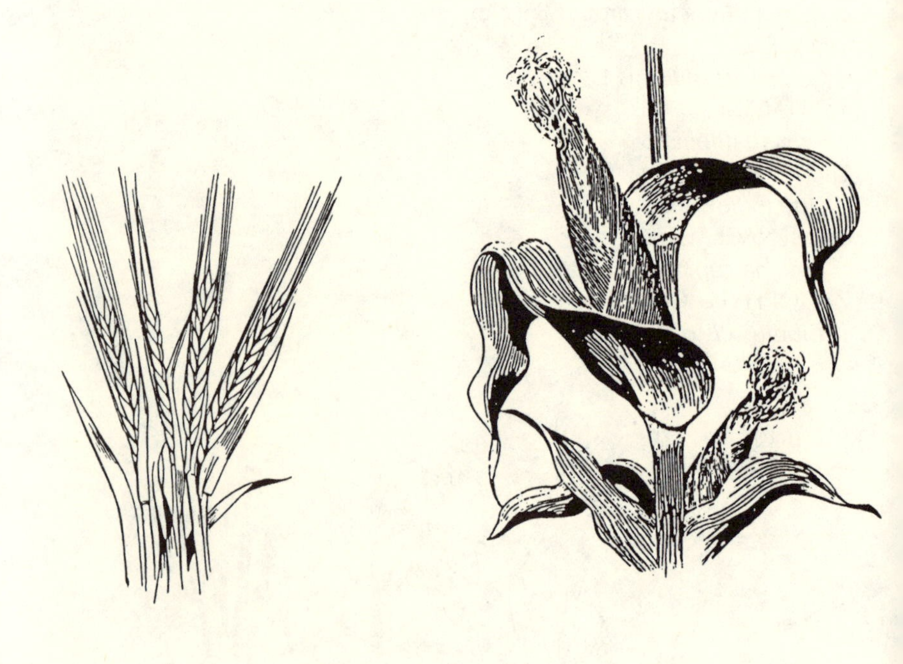

Etymology of the Scientific Names of CROPS

THERE are several lists that one could make of the world's most important crops. In addition to bias of the maker, such lists depend on whether importance is defined in terms of land area devoted to a crop, economic return to growers, or the number of people dependent on a crop for some of their daily sustenance. This is not the place to debate how such a list should be constructed or what crops should be included. However, there are some crops that are on all lists: wheat, rice, and maize meet all three criteria. Other crops that should be included in any complete list are root crops such as the potato, tree crops such as the coconut, and legumes such as the soybean. The crops herein do not include all crops of interest to humans but do include most of those that we depend upon for life. My interest has been etymological rather than agronomic; therefore, I have included some crops because of their name rather than their worldwide importance.

Of the 35 crops, 11 are grasses and 7 are legumes, which reflects the importance of these two families as sources of the world's food. There are 32 genera and 15 families included and, for most, the etymology is clearer than it is for weeds. Only 3 are named after people (cowpea, mungbean, and tea), while 7 are named after places. Eight names have an obscure origin, which makes description easy but unsatisfying. Twenty-eight were originally named by Linnaeus and 10 of these were subsequently modified; therefore, only 7 species in this group were not named by Linnaeus.

Ananas comosus **(L.)Merr.** Pineapple

(A′-na-nas co′-mo-sus)

ANANAS: (S) *ananas* = pineapple
ANNONA: (L) victuals; a year's harvest
ANNUA: (L) a year
KOMMOS: (G) ornamentation
COMOSUS: (L) bearing a tuft of hairs or leaves; with many leaves

The Spanish origin of the generic name is probably from *annona,* the Latin word for a year's harvest, which may have been Linnaeus's allusion to the long growing season of pineapple. The specific name aptly describes pineapples, which have abundant, thick, sharp-pointed, spiny-edged leaves that form the base of the plant. The multiple fruit, formed by fusion of about 100 flowers and bearing a tuft of leaves on its top, can be used to vegetatively reproduce pineapples.

Arachis hypogaea **L.** Peanut or Groundnut

(A-rack′-iss hi-po-ga-e′-uh)

ARACHIS: (G) perhaps modification of *arakis* = chickling vetch; dim. of *arakos,* name of a leguminous plant
HYPO: (G) below, under, down
HYPOGENOUS: (G) growing on the lower side
GAIA: (G) the earth, land

Origin of the generic name for peanut, as it is called in the United States, or groundnut, as it is called in most of the rest of the world, is not clear. It produces nuts underground (*gaia*) in hypogenous fashion, and the specific name is appropriate although not immediately intelligible to those not familiar with botanical jargon.

Avena sativa **L.** Oats

(Ah′-v-na sa-t′-vah)

AVENA: (L) oats
SATUS: (L) a planting
SATIVUS: (L) sown, planted; cultivated, not tame or domesticated

The entire scientific name alludes to the tame, domesticated, or planted oat as opposed to the wild oat, *A. fatua.*

***Beta vulgaris* L.** Sugarbeet and Red Beet

(Bay′-ta vul-gare′-iss)

BETA: (L) beet
VULGATUS: (L) ordinary, common
VULGARE: (L) to make known
VULGUS: (L) mob, common people

The sugarbeet and table beet, or red beet, share this scientific name and are closely related. They both were regarded as common or ordinary plants when named. It has been suggested that the generic name was assigned because of the resemblance of the seed of some species to the second letter (beta) of the Greek alphabet.

***Cajanus cajan* (L.)Huth** Pigeon Pea

(Ca-jan′-us ca-jan′)

CAJAN: (L) prob. fr. Malay *kachang* = bean or pea; also pigeon pea

The generic and specific names have the same origin, which is unique among the plants included herein. The names are not descriptive of looks or behavior, but they do reveal what it is in the language from which the name originated. Pigeon peas are cultivated for their edible seeds in India and other semiarid tropical areas.

***Camellia sinensis* (L.)Ktze.** Tea

(Ca-meal′-e-ah sih-nen′-sis)

CHAMAI: (G) on the ground
KAMEL: after G. J. Kamel, a Moravian Jesuit missionary, died 1706
SIN: (L) fr. China
ENSIS: (L) suffix meaning place of origin

Georg Joseph Kamel was a Moravian Jesuit missionary who first described tea and brought it from the Orient to London. Camellia is the Italian version of his surname. This small tree with lanceolate leaves and fragrant white flowers is maintained as a shrub by careful pruning; it can grow up to 15 m. Linnaeus originally placed it in the genus *Thea*. It is a native of China and has been cultivated there for 2000–3000 years. It is now grown in many subtropical regions and in mountainous tropical regions.

Cicer arietinum L. Chickpea
(Sigh′-sir r-e-eh-tin′-um)

CICER: (L) chickpea
KRIOS: (G) a kind of pea
ARIETIS: (L) a ram
INUS: (L) suffix meaning belonging to

This member of the legume family would be recognized as a pea by those who know the garden pea (*Pisum sativum* L.). It is widely cultivated for food in India and parts of Europe. The specific name is derived from the curved seed pod, which, with some micro-imagination, resembles a ram's horn.

Cocos nucifera L. Coconut
(Co′-cose new-sif′-er-uh)

COCOS: possible origin fr. Spanish or Portuguese *coco* = bogeyman
NUC: (L) nut
FERA: (L) fr. *ferre* = to bear or carry

The origin of the generic name is obscure but is related to the Spanish word *coco,* which literally means bogeyman. It is a reference to baby talk co! boo!), which apparently is related to a resemblance of a dehulled coconut to a grotesque head. The name could also have come from *macaco,* the Portuguese word for monkey, because the dehulled coconut resembles a monkey's face when viewed from the end where the shoot emerges from one of three depressions in the endocarp. The specific name is completely appropriate for this nut-bearing member of the palm family (Arecaceae).

Coffea arabica L. Coffee
(Coff′-e-ah uh-rab′-ih-cah)

QAHWAH: (A) coffee
ARABICA: (L) fr. Arabia

Coffea canephora Pierre ex Froehner Coffee
(Coff′-e-ah ca-nef′-or-uh)

COFFEA: see above
CANNA: (L) reed or cane
KANNA: (G) reed; reed mat
PHERO: (G) to bear or carry

C. arabica is the dominant species and an upland crop, which does not do well in humid tropical lowlands. *C. canephora* is cheaper to produce and has less specific environmental requirements. It is regarded as of lower quality and is used extensively in instant coffee blends. *C. arabica* is often but not always grown in shade and under cooler conditions. Optimal environments occur near the equator at 1500–2500 m with a temperature of 15–25° C. The coffee plant is an evergreen and its fruit is borne on the canes or branches. Although most is grown in South America, Africa, and Asia, more is consumed in the United States and Europe.

***Elaeis guineensis* Jacq.** Oil Palm

(L′-a-iss ginn-e-n′-sis)

ELAION: (G) oil
ELAIA: (G) an olive
GUINE: (L) fr. Guinea
ENSIS: (L) suffix meaning place of origin

The scientific name for the oil palm is quite precise. The oil palm is a native of the tropics and came from Guinea, formerly a large region of West Africa; it is now divided into several countries.

***Glycine max* (L.)Merr.** Soybean

(Gly′-scene max)

GLYKYS: (G) sweet
MAX: (L) large
MAXIMUS: (L) greatest, largest

The soybean is known for its high protein and fat content, not for its sweet taste. The grain is edible whole and has a nutlike taste; it does not taste especially sweet. This is one of the largest members of the genus.

***Gossypium hirsutum* L.** Cotton

(Goss-sip′-e-um hear′-soo-tum)

GOSSYPION: (L) the cotton tree
QOTHN: (A) a soft substance
HIRTUS: (L) rough; hairy
UTUM: (L) suffix meaning provided with

The cotton plant is not a tree nor is it especially hairy. The specific name is derived from the hairs (or cotton) that form the epidermis of cotton seeds. It is not possible to trace the origin of the generic name beyond Latin.

Hevea brasiliensis (Willd. ex A.Juss.) Muell.-Arg. Rubber Tree

(Heh-v′-uh bra-sil-e-n′-sis)

HEVEA: (S) *jebe* = rubber plant, prob. of American Indian origin
BRASILI: (L) fr. Brazil
ENSIS: (L) suffix meaning place of origin

This native of the Amazon basin tropical rain forests has a generic name of Spanish origin. Today, more than 90% of the world's natural rubber is produced in southern and eastern Asia, with very little being produced in Brazil. The fast-growing, deciduous tree grows 25 m tall.

Hordeum vulgare L. Barley

(Hor′-d-um vul′-gare)

HORDEUM: (L) barley
HORRERE: (L) to bristle
VULGATUS: (L) ordinary, common
VULGARE: (L) to make known
VULGUS: (L) mob, common people

The Latin name for barley is the generic name for this common, small-grain plant. The specific name was derived from its common association with humans and their dwelling places.

Ipomoea batatas (L.)Lam. Sweet Potato

(Ip-o-me′-ah ba-ta′-tahs)

IPS: (G) a worm; may also mean a bindweed
IPOS: (G) a worm that eats vines or wood
OMOEA: (G) *homoios* = like
BATATA: (S) *patata* var. of *batata* = sweet potato; prob. fr. Taino word for potato

The Tainoiis, an extinct tribe of Indian aborigines of the Greater Antilles, Bahamas, and Haiti in the Caribbean, gave the name to this viney member of the Convolvulaceae. It resembles weedy members of the genus in its spreading growth pattern, but it is valued for its tuber production and foliage; it is picked and cooked as a vegetable.

Lycopersicon esculentum **Mill.** Tomato

(Li-co-per′-sih-con si-q-len′-tum)

LYKOS: (G) wolf; a hood; or door knocker
LYCOPERSION: (L) an Egyptian plant
PERSICUM: (L) peach
PERSICUM: (L) fr. Persia
ESCULENT: (L) edible, good to eat
ULENTUS: (L) a suffix meaning abundance or marked development
ESCA: (L) food
EDERE: (L) to eat

The etymology of wolf peach is given by other authors but does not make obvious sense. Club mosses belong to the genus *Lycopodium,* which means wolf foot and does make sense because of claw-shaped roots. One does not immediately think of a claw when observing tomato roots but there may be a resemblance. The fruit, while not peachlike, has the same shape although not the same texture, color, or taste. It stretches the most fertile modern botanical imagination to see a wolf peach, a hood, or door knocker in a tomato plant. The Latin derivation makes some sense except the tomato is thought to be native of the New World because all related wild species are natives of the Andean region. Mexico is the most probable region of domestication, and it was cultivated there before being taken to Europe in the early 1500s. Thus, it is unlikely that it originated in Egypt or Persia. The tomato is edible but for centuries was thought to be poisonous, probably because of its unfortunate familial association with known poisonous species such as black nightshade (*Solanum nigrum* L.), mandrake (*Mandragora officinarum* L.), and belladonna (*Atropa bella-donna* L.). References to eating tomatoes are rare prior to the mid-1800s, although Thomas Jefferson, who was ahead of his time in so many ways, wrote of cultivating them in Virginia in 1782.

Manihot esculenta **Crantz** Cassava, Manioc, Yuca

(Man′-e-oat s′-q-len′-ta)

MANIHOT: (F) cassava; prob. of Tupian indian origin
ESCULENT: (L) edible, good to eat
ULENTUS: (L) suffix meaning abundance or marked developement
ESCA: (L) food
EDERE: (L) to eat

Most Americans will know this crop, if they know it at all, as the main ingredient in tapioca pudding. Because of its high carbohydrate content, it is an important source of calories to many of the world's people in Africa and South America. The Tupians, a tribe of Brazilian indians, are the likely origin of the name for the plant that is harvested as a large edible root. Cassava was probably first domesticated in southern Mexico, Guatemala, and northeastern Brazil. It is a short-lived perennial that grows 0.5–1.3 m tall and bears tubers that may weigh 15–20 kg and be up to 1 m long.

***Musa paradisiaca* L. var. sapientum (L.)Ktze.** Banana

(Moo′-sa pair-ah-d-c-ah′-cuh)

MUSA: (A) derived from *mouz* or *mawzah* = banana
PARADISIACA: (L) of or related to paradise
SAPIENS: (L) knowing
SAPERE: (L) to know

Musa may have been derived from the Arabic word for banana or from the Muses, the patrons of the arts. Linnaeus said he derived it from the surname of Antonius Musa (63–14 B.C.) a physician to Octavius Augustus, the first Roman emperor. The banana was unknown in the Mediterrenean region in classical times but accounts reached the region through conquests and explorations of Alexander the Great (356–323 B.C.). It was known to the Arabs and appears in the Koran as the tree of paradise, which is equivalent to the Christian tree of knowledge. The Arabs regarded it as the tree of paradise because of its ability to bear several crops of a desirable fruit, and perhaps also because of its occurrence near sources of water, so important to desert life.

***Oryza sativa* L.** Rice

(Or-i′-za sah′-t-va)

ORIZA: (G) rice
ORYZA: (L) rice
SATIVA: (L) sown
SATIVUS: (L) that which is sown, planted, cultivated
SATUS: (L) a planting
SERO: (L) to sow

The Latin word for rice has been retained as the generic name and no further derivation is possible. Rice is cultivated around the world and probably one of every three of the world's people depends on it for a major portion of their daily sustenance, more than for any other crop. There are over 100,000 rice varieties in the world.

***Pennisetum americanum* (L.)Leeke** Pearl Millet

(Peh-nih′-c-tum ah-mare-ih-ca′-num)

PENNI: (L) feather, wing
SETUM: (L) fr. *seta* = bristle
AMERICAN: (L) fr. America
UM: (L) suffix indicating possession

The generic name refers to the terminal, dense, compact, usually cylindrical spike inflorescence. In many species, each spikelet is surrounded by bristles giving the appearance of a bristly wing or feather. This genus also contains several weedy species. It is American in origin but is an important human food crop in India and much of Africa.

***Phaseolus vulgaris* L.** Bean

(Fay′-z-o-lus vul-gare′-iss)

PHASEOLUS: (L) dim. of *phaseolos* = bean
PHASELOS: (G) a little boat
VULGATUS: (L) ordinary, common
VULGARE: (L) to make known
VULGUS: (L) mob, common people

This is the common bean with many modern variations, including the kidney, string, and bush beans of farm and garden. The name was probably derived from *phaselos,* Greek for a little boat, because the seed pod resembles a small boat.

***Saccharum officinarum* L.** Sugarcane

(Sack′-r-um o-fis-in-r′-um)

SAKCHAR: (G) *sakcharon* = sugar
SINGKARA: (M) sugarcane
ARON: (G) reed, a plant
ARUM: (L) fr. *arundin* = reed
OFFICINA: (L) an office
OPIFICINA: (L) originally a store-room or workshop, later pharmacy
UM: (L) suffix indicating possession

This large grass grows 3–4 m tall and is a tropical or semitropical crop that produces sugar in its stems or reeds. The specific name comes from the Latin for storeroom or office, later pharmacy. Its use implies that sugarcane or its products were once regarded as having medicinal value, and it may have appeared on an official (*officinale*) drug list and been dispensed by pharmacists. The sweet taste obtained from sucking or chewing the stem makes one believe it could have been included in the ancient pharmacy, even if only as a palliative.

***Secale cereale* L.** Rye

(C-cal′-a sear′-e-ahl-a)

SECALE: (L) rye
CEREALIS: (L) of Ceres, the goddess of grain
CRESCERE: (L) to grow

Ceres is the goddess of agriculture and all fruits of the earth in the Roman pantheon. The name can be literally translated as Mother Earth.

***Solanum tuberosum* L.** White or Irish Potato

(So-lay′-num too-ber-o′-sum)

SOLA: (L) sun
SOLOR: (L) to comfort, soothe
SOLARI: (L) to comfort or quiet
SOLAMEN: (L) a solace
SOLANUM: (L) literally means nightshade
NUM: (L) an interrogative particle usually implying a negative answer
TUMERE: (L) to swell
OSUM: (L) a suffix meaning full of, possessing

Solanum was probably assigned because deadly nightshade (*Atropa bella-donna*) was one of the first plants classified in this genus. Atropine is extracted from deadly nightshade, and legend tells us that witches brewed the potent juice from the plant in the dead of night, without sun, when witches work. The name could also have been derived from the Latin words for solace and comfort because of the narcotic or medicinal properties of some species. Tuber comes from *tumere,* the Latin for swelling, which tubers do; this species produces abundant tubers.

Sorghum bicolor **(L.)Moench** Sorghum

(Sor'-gum bi'-color)

SYRICUS: (L) fr. Syria
SURGUM: (L) great millet
GRANUM: (L) grain
BI: (L) two
COLOR: (L) color

The etymology of sorghum is not clear, but it is close to a combination of the Latin for Syria and grain, thus giving us a grain from Syria, which is a likely site of origin. It is also likely that it came directly from Latin. The specific name comes from the multicolored seed, which assumes several hues but is usually red to brown and tinged with purple.

Theobroma Cacao **L.** Cocoa

(Theo-bro'-ma ca-ca'-o)

THEO: (G) God
BROMA: (G) food
BORA: (G) food, meat
CACAIN (L) chocolate brown
NAHUATL: (S) cacao bean
CACAHUATL: (S) cacao bean

This is the source of cocoa and chocolate. Dried, usually fermented seeds of the cacao tree are used in preparations of cocoa, chocolate, and cocoa butter. It is a tree that bears fruit on its trunk and old branches. The flowers have a pink calyx and yellowish corolla, which is succeeded by fleshy yellow pods, each about 15 cm long and 7.5–10 cm in diameter.

Triticum aestivum **L.** Wheat

(Trih'-tih-cum s'-tih-vum)

TRITICUM: (L) wheat
TERERE: (L) to rub, to thresh, to grind
TRITUS: (L) rubbed
AESTUS: (L) summer
AESTIVUS: (L) pertaining to summer

Triticum durum Desf. Durum Wheat
(Trih′-tih-cum dur′-um)

TRITICUM: see above
DURUM: (L) fr. *durus* = hard

Wheat, like all small grains, has to be threshed (rubbed) to separate the grain from other plant parts. This process has been used since wheat has been used by humans and is the source of the generic name. *T. aestivum,* the dominant form, is bread wheat, which is grown as a spring- or fall-planted grain throughout the temperate zones of the world. Only rice (*Oryza sativa*) feeds more people, but wheat is grown on more acres than any other crop. *T. durum* is the hard wheat used for pasta and macaroni products and is grown mainly in Europe. The grain is usually, but not always, harder than soft or bread wheats.

Vigna radiata (L.)Wilczek Mung Bean
(Vig′-na ray-d-ah′-ta)

VIGNA: after Domenico Vigna, an Italian botanist, died 1647
RADIATUS: (L) to emit rays, rayed
RADIUS: (L) ray

Vigna unguiculata (L.)Walp. Cowpea
(Vig-na un-gwick′-u-la-ta)

VIGNA: see above
UNGUICULUS: (L) small nail, claw, talon, or hoof

Radiata refers to bean pods that radiate or spread out from a single axis on this erect, annual, early maturing bush or slightly vinelike herb. It is an important protein source in India and other tropical countries.

Unguiculata is usually used in botanical nomenclature to signify tapering to a claw or stalklike base. It often refers to petals, but here it probably refers to the long, linear, crescent, or coil-shaped (ram's horn) seed pods. Their length and general uncoiled shape resemble fingers with their nails. The name could also have been derived from the curved shape of the stamens. This legume has trifoliolate leaves with shapes ranging from spear-shaped to ovate.

Vitis vinifera L. Grape

(V'-tis vin-if'-er-uh)

VITI: (L) pertaining to the vine
VINIFER: (L) wine producing
VINUM: (L) wine
FERO: (L) to bear or carry

This scientific name is associated with what the plant produces not with its looks or behavior. It is a vine that produces grapes, which yield wine as the specific name indicates. More than 50% of the world's crop is produced in Europe, but grapes are grown on every continent. About 75% of the world's crop is made into wine or distilled beverages.

Zea mays L. Maize or Corn

(Z'-uh mayz)

ZEA: (L) single-grained wheat, a kind of grain
ZEIA: (G) a kind of grain
MAYS: (S) maize

Corn, as it is known in the United States, and maize, as most of the rest of the world calls it, is not a single-grain crop. The grain is produced in multiples on ears. However, each can be identified as a single grain on this tall, vigorous crop that is so important as an animal and human food crop. The United States is the world's largest producer.

Zingiber officinale Roscoe Ginger

(Zin-guy'-ber o-fis-in-al')

ZINGIBER: (L) ginger
ZINGIBERI: (G) ginger; prob. fr. Sanskrit *srngavera* = ginger
OPIFICINA: (L) originally a store-room or workshop, later pharmacy
OFFICINA: (L) an office

Ginger is a leafy, Asian or Polynesian perennial. The aromatic, tuberous rhizomes from which ginger is prepared are dried and ground to a fine powder. It is widely used as a spice and sometimes as a carminative, stimulant, or counterirritant medicine.

Zizania aquatica L. Annual Wild Rice
(Zih-zan′-e-uh ah-quah′-tih-ca)

ZIZANION: (G) wild grain; the tares of scripture
AQUATICA: (L) aquatic

A tall, annual grass with long leaves and large terminal panicles from which we obtain wild rice. Many weed scientists regard it as a weed, but many farmers grow it for a good profit. It is worth noting that *zizanion* has been proposed as an important root of the term weed and zizaniology was suggested as a name for the study of weeds (Antonopoulos, A. 1976. Abst. 131 Annu. Meet. Weed Sci. Soc. Am., Denver, Colo.)

Scientific Names of Crops by Plant Family

ARECACEAE
Cocos nucifera L.
Elaeis guineensis Jacq.
BROMELIACEAE
Ananas comosus (L.)Merr.
CHENOPODIACEAE
Beta vulgaris L.
CONVOLVULACEAE
Ipomoea batatas (L.)Lam.
EUPHORBIACEAE
Hevea brasiliensis (Willd. ex A. Juss.)Muell.-Arg.
Manihot esculenta Crantz
FABACEAE
Arachis hypogaea L.
Cajanus cajan (L.)Huth
Cicer arietinum L.
Glycine max (L.)Merr.
Phaseolus vulgaris L.
Virgna radiata (L.)Wilczek
Vigna unguiculata (L.)Walp.
MALVACEAE
Gossypium hirsutum L.
MUSACEAE
Musa paradisiaca L. var. *sapientum* (L.)Ktze.
POACEAE
Avena sativa L.
Hordeum vulgare L.
Oryza sativa L.
Pennisetum americanum (L.)Leeke
Saccharum officinarum L.
Secale cereale L.
Sorghum bicolor (L.)Moench
Triticum aestivum (L.)
Triticum durum Desf.
Zea mays L.
Zizania aquatica L.

RUBIACEAE
Coffea arabica L.
Coffea canephora Pierre ex Froehner
SOLANACEAE
Lycopersicon esculentum Mill.
Solanum tuberosum L.
STERCULIACEAE
Theobroma cacao L.
THEACEAE
Camellia sinensis (L.)Ktze.
VITACEAE
Vitis vinifera L.
ZINGIBERACEAE
Zingiber officinale Roscoe

GLOSSARY

Achene: A hard, dry, indehiscent, one-seeded fruit with a single cavity. The pericarp is free from the seed.

Angiosperm: A flowering plant characterized by having seeds enclosed in an ovary.

Anther: The pollen-bearing part of the stamen.

Auricle: A claw or earlike appendage, frequently appearing at the junction of the leaf sheath and blade in grasses.

Awn: A stiff or hard, bristlelike appendage, usually at the end of an organ.

Axil: The upper angle between a leaf or a branch and the stem or axis that bears it.

Biennial: Living for two years.

Bulb: A short, basal, mostly underground stem surrounded by thick, fleshy leaf blades.

Bract: A reduced or modified leaf, particularly the scalelike leaves at the base of a flower cluster.

Calyx: The outer whorl of a flower made up of united or divided sepals. It is commonly green and leaflike but may resemble petals.

Capsule: A dry dehiscent fruit originating from two or more carpels.

Carpel: One of the units of a pistil. If a pistil has one carpel, it is simple; if it has more than one, it is compound.

Cordulate: Heart shaped with the point upward.

Corolla: The inner whorl of the flower made up of united or divided petals. Usually the showy part of the flower.

Corymb: A flat-topped inflorescence with the lower branches longer than the upper so all flowers are at the same level. The outer flowers usually open before the inner ones.

Culm: The aboveground stem of grasses or grasslike plants.

Cyme (cymose, adj.): A broad, more or less flat-topped inflorescence in which the main axis terminates in a single flower that opens before the lateral flowers arising beneath.

Decumbent: Lying flat with the apex ascending. Used with reference to stems.

Dentate: Having marginal teeth pointed outward but not forward. (See serrate)

Digitate: Fingered. A compound structure whose members arise and diverge from the same point, shaped like an open hand.

Dioecious: Having staminate (male) and pistillate (female) flowers on different plants.

Disk flower: One of the central tubular flowers of the composite flower head of the Asteraceae. The flower is a fleshy development of the receptacle about the base of the ovary.

Fern ally: A member of the fern family that is not leafy, as distinguished from the true leafy ferns. Ex. *Equisetum* sp.

Floret: One of the flowers in a dense inflorescence of small flowers. Commonly used to refer to a single flower in the complex inflorescence of the Poaceae (grasses) or the Asteraceae.

Glabrous: Smooth, without hairs.

Glaucous: Having a waxy bloom or white waxy or powdery coating.

Glomeruk: A small cluster of flowers in a compacted cyme.

Glume: One of a pair of dry bracts subtending and often enclosing the grass flower.

Hypogynous: Having the perianth and stamens inserted on the receptacle below the gynoecium.

Imbricated: Overlapping, like shingles on a roof.

Inflorescence: An aggregation of flowers on a plant. There are many different kinds of inflorescence, each representing a different shape and usually characteristic of a certain kind of plant.

Involucre: A group or ring of free or united bracts that subtend or enclose an inflorescence.

Labiate: Having the calyx or corolla so divided that one portion overlaps the other, lipped.

Lanceolate: Lance shaped, much longer than broad, arising from a broad base and tapering toward the apex.

Leaf sheath: The basal part of a leaf that surrounds the stem. Common in grasses.

Lemma: The outer bract subtending the grass floret. (See palea)

Nectary: A gland or tissue for secreting nectar.

Node: The region (usually a swelling or joint) on a stem from which a leaf or leaves arise, or on a root from which stems or roots arise.

Palea: The upper bract subtending the grass floret. (See lemma)

Palmate: Having several parts radiating from a common axis, like the fingers from the palm of the hand.

Panicle: A compound inflorescence in which the main axis is branched one or several times.

Pappus: The chaffy, scaly, bristlelike, or plumose structure at the junction of the achene and the corolla in the Asteraceae. Often seen as a ring of hairs at the top of the fruit.

Pectinate: Having clefts or divisions so as to resemble a comb.

Pedicel: The stalk or stem of a flower in a flower cluster.

Peduncle: The stem of a solitary flower or the main stem of a flower cluster.

Perfect flower: A flower with stamens and pistils.

Perianth: Collectively, the calyx and the corolla. Often used when these are not differentiated.

Pericarp: The wall of the ripened ovary.

Perigynous: Having the perianth and stamens inserted on the receptacle around the gynoecium, or where the ovary is partly embedded in the receptacle.

Petiole: The stem or stalk of a leaf.

Pinnate: Having a common axis with segments arranged opposite or alternate on either side.

Pistil: The female part of the flower consisting of the stigma, style, and ovary.

Plumose: Hairy with side hairs along the main axis like the plume of a feather.

Raceme (racemose, adj.): An inflorescence with flowers arising on pedicels from a common central axis with the youngest toward the tip.

Rachis: The axis of a spike or raceme or compound leaf.

Ray flower: One of the outer or marginal flowers of the composite flower head of the Asteraceae. Ray flowers are ligulate or strap shaped in contrast to the tubular shape of the disk flowers.

Receptacle: The more or less enlarged end or axis of a flower stalk that bears the flower parts.

Rhizome: A horizontal underground stem, distinguished from the root by the presence of buds and scales.

Rootstock: A horizontal underground stem bearing roots and aerial stems along its axis or from its tip.

Sedge: Any of a family (Cyperaceae) of usually tufted, wetland plants (although they do grow on drylands). They differ from related grasses in that they have achenes and solid stems that are triangular in cross section. There are three rows of narrow, pointed leaves and minute flowers borne in spikelets.

Sepal: One of the segments of the calyx (usually green).

Serrate: Having marginal teeth pointed forward. (See dentate)

Sessile: Without a stalk, stem, or petiole. Joined directly at the base.

Spike: An unbranched, elongated inflorescence with sessile or nearly sessile flowers.

Squarrose: Having parts or processes (usually tips) spreading or recurved.

Spikelet: The segment of the grass inflorescence enclosed by a pair of glumes.

Stamen: The male, pollen-bearing organ, consisting of anther and stalk or filament.

Staminate: Having stamens.

Stipule: An appendage, often leaflike, occurring at the base of a leaf or at the node of a stem.

Stolon: A modified aboveground stem that creeps and can root from its nodes.

Style: A short or long, branched or unbranched stalk arising from the ovary and bearing the stigmas or stigma.

Umbel: An inflorescence of few to many flowers on stalks of about equal length arising from the top of a peduncle.

Utricle: Usually a one-seeded, indehiscent fruit with a thin, bladdery, persistent ovary wall.

Trifoliolate: Having three leaflets.

Tuber: An enlarged, fleshy, short, thickened underground stem.

Whorl: The structure or geometry represented by a ring of similar organs arising from a single node.

SELECTED REFERENCES

Alcock, Randal H. 1876/1971. *Botanical Names for English Readers.* London: L. Reeve; Detroit: Grand River Books. 236 pp.

Aldrich, R. J. 1984. *Weed-Crop Ecology: Principles in Weed Management.* N. Scituate, Mass.: Breton Publ. 465 pp.

Bailey, L. H. 1933/1963. *How Plants Get Their Names.* N.Y.: Macmillan; N.Y.: Dover Publ. 181 pp.

Bailey, L. H., and E. Z. Bailey. 1941. *Hortus the Second.* N.Y.: Macmillan. 778 pp.

Baker, H. G. 1965. Characteristics and modes of origin of weeds, 147–72. In *Genetics of Colonizing Species,* ed. H. G. Baker. N. Y.: Academic Press.

Beste, C. E. 1983. *Herbicide Handbook of the Weed Science Society of America,* 5th ed. Champaign, Ill.: Weed Sci. Soc. of America. p. xxiv.

Blatchley, W. S. 1912. *The Indiana Weed Book.* Indianapolis, Ind.: Nature Publ. 191 pp.

Borrer, D. J. 1960. *Dictionary of Word Roots and Combining Forms, Compiled From the Greek, Latin, and Other Languages, with Special Reference to Biological Terms and Scientific Names.* Palo Alto, Calif.: W-P Publ. 134 pp.

Brenchley, W. E. 1920. *Weeds of Farm Land.* London: Longmans, Green. 239 pp.

Brown, Roland Wilbur. 1954. *Composition of Scientific Words: A Manual of Methods and a Lexicon of Materials for the Practice of Logotechnics.* Washington, D.C.: Author. 832 pp.

Bunting, A. H. 1960. Some reflections on the ecology of weeds, 11–26. In *The Biology of Weeds,* ed. J. L. Harper. Oxford, Eng.: Blackwell Sci. Publ.

Clute, Willard N. 1942. *The Common Names of Plants and Their Meaning.* Indianapolis, Ind.: Willard N. Clute. 164 pp.

Composite list of weeds. 1984. *Weed Sci.* 32(Suppl. 2):1–37.

Cunningham, John J., and Rosalie J. Cote. 1977. *Common Plants—Botanical and Colloquial Nomenclature.* N.Y.: Garland Publ. 120 pp.

Emerson, R. W. 1876. Fortune of the republic, 509–44. In *Miscellanies,* Vol. 11, *The Complete Works of Ralph Waldo Emerson.* N.Y.: Houghton Mifflin.

Gledhill, D. 1985. *The Names of Plants.* Cambridge, Eng.: Cambridge Univ. Press. 159 pp.

Gould, Sydney W., and Dorothy C. Noyce. 1965. *Authors of Plant Genera,* Vol. 2, *International Plant Index.* N.Y.: Botanical Garden, N.Y. Agric. Exp. Sta.; New Haven: Conn. Agric. Exp. Sta. 336 pp.

Gove, Phillip B., ed. 1961. *Webster's Third New International Dictionary of the English Language.* Springfield, Mass.: Merriam. 2662 pp.

Hatfield, Audrey Wyne. 1969. *How to Enjoy Your Weeds.* N.Y.: Collier Books. Div. of Macmillan. 192 pp.

Holm, L., D. L. Plucknett, J. V. Pancho, and J. P. Herberger. 1977. *The World's Worst Weeds: Distribution and Biology.* Honolulu: Univ. Press of Hawaii. 609 pp.

Jaeger, Edmund C. 1959. *A Source Book of Biological Names and Terms,* 3d ed., rev. 2d printing. Springfield, Ill.: Charles C. Thomas. 319 pp.

Johnson, A. T. 1931/1971. *Plant Names Simplified—Their Meanings and Pronunciation,* 2d ed. London: W. H. and L. Collingridge Ltd.; Detroit: Grand River Books. 146 pp.

King, Lawrence J. 1966. An introduction to weeds, 1–32. In *Weeds of the World: Biology and Control.* N.Y.: Interscience Publ.

Linnaeus, C. 1938. *The Critica Botanica of Linnaeus.* Trans. Sir Arthur Hort. London: The Ray Soc. 239 pp.

Little, W., H. W. Fowler, and J. Coulson. 1973. *The Shorter Oxford English Dictionary on Historical Principles,* 3d ed. 2 vols. Rev. and ed. C. T. Onions, Etymologies rev. G. W. S. Friedrichsen. Oxford, Eng.: Clarendon Press. 2672 pp.

Lowell, James Russell. 1890. A fable for critics (1838), 117–48. In *Lowell's Complete Poetical Works.* Boston: Houghton and Mifflin.

Muenscher, W. C. 1960. *Weeds,* 2d ed. N.Y.: Macmillan. 560 pp.

Plowden, C. Chicheley. 1970. *A Manual of Plant Names,* 2d ed. N.Y.: Philosophical Library. 260 pp.

Radford, Laurie Stewart. 1974. Botanical names, 57–78. In *Vascular Plant Systematics,* ed. Albert E. Radford, William C. Dickinson, Jimmy R. Massey, and C. Ritchie Bell. N.Y.: Harper & Row.

Robinson, Benjamin L., and Merritt L. Fernald. 1908. *Gray's New Manual of Botany: A Handbook of the Flowering Plants and Ferns of the Central and Northeastern United States and Canada,* 7th ed. N.Y.: American Book Co. 926 pp.

Shakespeare, William. 1597. *Richard II,* Act 3, Scene 4.

Smith, A. W. 1963. *A Gardener's Book of Plant Names: A Handbook of the Meaning and Origins of Plant Names.* N.Y.: Harper & Row. 428 pp.

Smith, A. W. 1972. *A Gardener's Dictionary of Plant Names,* rev. ed. W. J. Stearn. N.Y.: St. Martin's Press. 391 pp.

Spencer, Edwin Rollin. 1940. *Just Weeds.* N.Y.: Scribner's Sons. 333 pp.

Stearn, William T. 1973. *Botanical Latin—History, Grammar, Syntax, Terminology and Vocabulary,* 2d ed. Newton Abbot, Devon, Eng.: David and Charles (Holdings) Ltd. 566 pp.

Tennyson's Poems. 1878. Boston: James R. Osgood. 361 pp.

Thomas, W. L., Jr., ed. 1956. *Man's Role in Changing the Face of the Earth.* An international symposium under the co-chairmanship of C. Sauer, M. Bates, and L. Mumford. Sponsored by the Wenner-Gren Foundation for Anthropological Research. Univ. of Chicago Press, Chicago.

INDEX